Diognei de Matos

VLC Communication in Smart Cities

Diognei de Matos

VLC Communication in Smart Cities

Development of a VLC communication applied to I2V systems in emerging smart cities

ScienciaScripts

Imprint

Any brand names and product names mentioned in this book are subject to trademark, brand or patent protection and are trademarks or registered trademarks of their respective holders. The use of brand names, product names, common names, trade names, product descriptions etc. even without a particular marking in this work is in no way to be construed to mean that such names may be regarded as unrestricted in respect of trademark and brand protection legislation and could thus be used by anyone.

Cover image: www.ingimage.com

This book is a translation from the original published under ISBN 978-613-9-74030-7.

Publisher:
Sciencia Scripts
is a trademark of
Dodo Books Indian Ocean Ltd. and OmniScriptum S.R.L publishing group

120 High Road, East Finchley, London, N2 9ED, United Kingdom
Str. Armeneasca 28/1, office 1, Chisinau MD-2012, Republic of Moldova, Europe
Printed at: see last page
ISBN: 978-620-6-24857-6

Summary

Urban mobility plays an important role in emerging smart cities, addressing issues such as congestion, pollution and especially accident reduction. The use of new technologies such as VLC (Visible Light Communication) could contribute in different aspects of infrastructure-to-vehicle (I2V) communication such as traffic lights and streetlights with potentials to become smart. One of the major challenges related to VLC rests on *flickering*, interference and scintillation. This paper seeks to present a new VLC communication technique to enhance and make smart the current traffic lights and poles by enabling them to transmit information to vehicles using the existing structures. For this, a prototype using microcontrollers and LEDs was developed. Results obtained demonstrate the feasibility of the proposed solution for I2V communication in smart cities.

Keywords: VLC, I2V, Smart Traffic Light, Smart Post.

Summary

CHAPTER 1

Introduction

1.1   Motivation and Context

The search for more energy-efficient and cheaper technologies has become increasingly important in recent times. Creating a technology that is able to solve a problem is not enough, it is also necessary that the technology has an affordable cost, satisfactory energy efficiency and the most efficient use of accessible resources in the environment where it will be applied. In the context of emerging smart cities, urban mobility is becoming increasingly complex, requiring more precise devices and systems that provide faster and more effective responses. Therefore, the use of new technologies has become necessary and present, opening possibilities for alternative strategies to solve specific problems. A good strategy for communication, which has been gaining space in research, is communication by visible light, in English *Visible Light Communication - VLC*, which consists of transmitting data through light, without harming its normal use.

VLC technology can provide significant advances for intelligent transportation systems (ITS) (KUMAR et al., 2012). From the point of view of smart traffic lights, since they will be able to transmit their state to vehicles, incorporating such functionality into Advanced Driver Assistance Systems (ADAS), these will serve as an aid and alert to the driver or even serve as a source of information for the decision making of recent autonomous vehicles (KUMAR; NERO; AGUIAR, 2007). Following the same reasoning for smart poles, since they have the ability to transmit information to the vehicle, it is possible to develop accurate localisation aid systems on two fronts. The first is characterised by the georeferenced mapping of the poles associated with an identifier. Thus, when the vehicle passes a particular pole, its identification will enable the precise localisation of this vehicle on the map. The second front is characterised by the estimation of the relative distance of the vehicle to the pole as a function of the light intensity received by it, thus performing trilateration, given that at any given time the vehicle will be in the field of view of possibly three different light sources. Therefore, the development of techniques using VLC for smart cities is indispensable, motivated mainly by the issue of efficiency regarding energy consumption and reuse of existing resources, thus reducing the cost of implementing these technologies with the reuse of resources already available (YAQOOB et al., 2017).

VLC technology has already been studied by several researchers and used in many applications. According to Pohlmann (2010), the use of visible light to transmit data has several advantages and avoids problems with transmission through electromagnetic waves outside the visible spectrum, such as health-related problems, since moderate exposure to light is considered safe for the human body, among other problems such as interference by electromagnetic radiation. Pohlmann (2010) states that light can be used in hospitals and other institutions without worrying about these problems. Other advantages are also highlighted such as the fact that visible light is free of charge, as there are no property rights over visible light, eliminating the payment of fees. VLC technology is safe, especially when used indoors, as it does not exit through walls, making it difficult to steal data as can happen with Wi-Fi and other technologies (POHLMANN, 2010).

All over the world, technology solutions for cities are being created by small businesses and multinationals as well as governments. "Cities have never been more populous. Two hundred years ago only three cities in the world (London, Tokyo and Beijing) had more than one million inhabitants", says Tonon (2013). With the agglomeration of the population, several problems arise, which makes it necessary to think of smart cities capable of contributing to a more sustainable control of resources. According to Tonon (2013), the concept of smart *cities* is defined by the use of technologies that seek to improve urban infrastructure and increase the efficiency of urban centres, improving the quality of life in them. This idea has gained momentum in recent years and there are already many smart cities such as Songdo in South Korea and Masdar in Dubai. Tonon (2013) states that the investment that is in common when it comes to smart cities always focuses on five main areas: environment, mobility, citizen-government interaction, quality of life and economy/creative people.

In this work, a technique using VLC is proposed to provide current LED traffic lights and streetlights with the ability to transmit data by light, thus contributing to ITS in the context of I2V communication. To this end, an electronic system with microcontrollers and LEDs was developed, as well as a communication protocol that together seek to avoid interference caused by the external environment and transmit data correctly through light. In addition, the prototype developed in this work can be used in other applications for transmitting messages by light, and can be used to transmit advertisements and other information through the lampposts in different situations.

1.2 Objectives

1.2.1 General Objective

Develop a VLC communication system for I2V systems on poles and traffic lights in smart cities.

1.2.2 Specific Objectives

- Develop a unidirectional serial communication protocol through a single communication channel which is light.

- Develop an ID-based location system based on pole positions that can be recognised by an intelligent vehicle.

- Develop a smart traffic light that communicates with the smart vehicle that will be able to receive the information transmitted by it.

- Validate the functioning of the system by simulating the real situation in different situations with interferences from ambient lights and with the receiver on a moving device.

1.3 Justification

The justification of this project is given by the need to develop more efficient communication technologies regarding energy consumption and reuse of resources already existing in smart cities, thus reducing the cost of these technologies, considering that the light of the poles and the traffic light will already be available in the cities during the night, and can also be used during the day, mainly in closed places, but also in open places with more economical LED lamps.

CHAPTER 2

Conceptual Rationale

This chapter presents the bibliographic references studied for the development of the proposed project.

2.1 Smart Cities

2.1.1 Urbanisation

Living in large, populous cities is the future of humanity. People are increasingly choosing to live in big cities where access to the resources needed for a better quality of life is easier. According to Harrison and Donnelly (2011), "most of this urbanisation is self-motivated as mechanisation reduces the need for manual labour in agriculture and agricultural workers choose to move to cities in search of a better life". Population is a key method to reduce energy consumption and therefore CO_2 emissions (OWEN, 2009).

Harrison and Donnelly (2011) state that this transition to an urbanised population creates challenges for the planning, development and operation of cities that are stimulating new thinking in building professionals such as architects, urban planners, designers, engineers, utilities, social science, environmental, public finance and policy professionals, and recently, in information technology. Information systems have many roles to play in large cities.

2.1.2 Smart Cities Concept

The concept of smart cities is not new, it had its origins in smart growth in the late 1990s (BOLLIER, 1998), which advocated new urban planning policies. According to Caldwell (2002), Portland, Oregon, is widely recognised as an example of a Smart City. The concept has been adopted since 2005 by a number of technology companies for the application of complex information systems to integrate the operation of urban infrastructure and services such as buildings, transport, electrical and water distribution and public safety (HARRISON; DONNELLY, 2011). From there, the concept of smart cities has evolved to represent any type of technology-based innovation in the planning, development and operation of cities, for example, the deployment of electric vehicle services.

2.1.3 Information Technology in Smart Cities

The application of information technology in smart cities can produce several benefits, states Harrison and Donnelly (2011), who points out the following benefits:

- Reduce resource consumption, mainly energy and water, thus contributing to reductions in CO_2 emissions.

- Improve the utilisation of existing infrastructure capacity, therefore improving quality of life and reducing the need for traditional construction projects.

- Provide new services to citizens and passengers, such as real-time guidance on how best to operate multiple modes of transport.

- Improve commercial enterprise by publishing real-time data on the operation of municipal services.

- Reveal how energy, water and transport demands peak at a city scale so that administrators can collaborate to smooth these peaks and improve resilience.

Harrison and Donnelly (2011) state that these approaches have become viable as a result of recent progress in technology:

- The widespread use of digital sensors and digital control systems for the control and operation of urban infrastructure. Including traffic sensors, building management, digital utility meters, among others.

- The increasing penetration of fixed and wireless networks that allow such sensors and systems to be connected to distributed processing centres and, in turn, to these centres to exchange information with each other.

- The development of information management techniques, specifically standardising semantic models, which allow low-level information to be interpreted by processing centres and allow these centres to interpret each other in training.

- The development of computing power and new algorithms that allow these information flows to be analysed almost in real time in order to provide operational information, performance improvement, among other benefits.

These developments allow municipal governments to coordinate the operations of their multiple agencies in the same way as the commercialisation of medium and large companies operating since 1980 (HARRISON; DONNELLY, 2011).

2.2   Visible Light Communication - VLC

2.2.1   Definition of VLC

Visible light communication (VLC) is a type of optical wireless communication that involves electromagnetic waves in the visible spectrum. According to Truzzi (2016), VLC is a subset of *Optical Wireless Communication* (OWC), which utilises the visible light spectrum as a means of communication. VLC applications consider only the visible spectrum and not the infrared or ultraviolet spectrum

In general, commercial LEDs are used for communication in the VLC, as they have a broad white wave emission spectrum that covers all visible fields. But there are also situations where incandescent and other lamps are used. The advantage of LEDs is that they can be switched on and off quickly and can modulate data at high speed in visible light (POHLMANN, 2010). The use of visible light for data transmission has many advantages and eliminates drawbacks of transmission via electromagnetic waves outside the visible spectrum, such as health-related problems. Visible light can serve as a completely free infrastructure on which to base a networked communication complex.

Figure 1 shows the frequency spectrum, highlighting the visible light range. As can be seen, the frequency range of visible light goes from 430THz to 790THz.

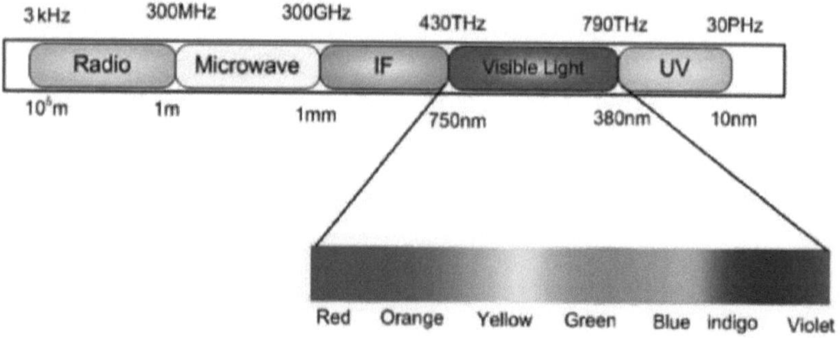

Figure 1 - VLC frequency spectrum

Source: Khan (2017)

According to (POHLMANN, 2010), VLC has generally been used in closed places where light does not come through the walls, contributing to information security, making it difficult to steal data as can occur with Wi-Fi. However, especially in situations where there is no need for such data protection, VLC can also be used in open locations.

When dealing with VLC, there are some characteristics that should be considered. Truzzi (2016) points out the following general characteristics:

- **Frequencies Involved**: Transmission systems use waves to communicate. In radio and optical communications, electromagnetic waves are used to carry the information. The electromagnetic spectrum is very broad and is categorised by amplitude or frequency.

- **Flicker**: This is the rapid variation of light intensity. In VLC, this characteristic is always present in the transmitter, because the transmission method uses intensity variations to transmit the signal. In addition, it is important that the light variation is at a very high frequency that cannot be perceived by the human eye.

- **Intensity Modulation - Direct Detection (IM-DD)**: Two methods of VLC communication are generally possible. The first is coherent transmission/detection, which is usually associated with laser diodes (LD). The second is IMDD which is associated with LEDs. In IMDD the intensity of the optical source is modulated by the signal, and demodulation is done by direct detection of the carrier optics and conversion is done using a photo-detector. According to Truzzi (2016), IMDD VLC systems are cheaper and easier to implement and the laser can be dangerous for the human eye. so IMDD is more convenient for daily commercial applications.

- **Signal to noise ratio (SNR)**: It is the ratio of signal power to noise power. It is a dimensionless parameter, very important for every communication system. It serves as a quality factor for the system. If the SNR is low, the transmission will not be done properly, it is recommended to reach very high values of SNR, which increases the cost of the system, which will consume more energy and the system may also become more complex.

- **Propagation and multipath reduction**: VLC signals propagate in open space. The signal arriving at the receiver has two components: The direct component and the diffusive component. The diffusive component is due to the reflected part of the signal that is generated by the interaction with the obstacle. The multipath propagation is due to the diffusive component of the signal. A consequence of this propagation is signal attenuation, which can also occur due to shadowing from obstacles between the source and the receiver.

- **Peak-to-Average Ratio (PAPR)**: This is the ratio of the square of the peak to the variance of the wave. This ratio is used to measure how "big" the waveform is.

2.2.2   VLC architectures

A VLC system consists of two parts, the transmitter and the receiver. Both parts usually consist of three common layers. These layers are the physical layer, the MAC (*Medium Access Control) layer,* and the application layer. Figure 2 shows a reference model for VLC communication systems with these layers, according to Khan (2017).

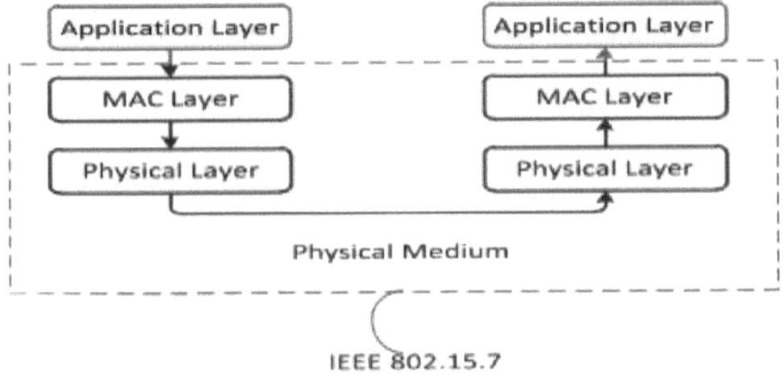

Figure 2 - VLC system layers

Source: Khan (2017)

- **Application Layer**: This layer contains the system responsible for using the information, managing it and providing an interface so that the user can use and manipulate it, and can also transmit it to other systems.

- **MAC Layer**: It is the medium access control layer, and performs several actions such as mobility support, mitigation, visibility and security, flicker mitigation schemes, colour function support, among others. The topologies supported by this layer are *peer-to-peer, broadcast* and star. Star communication is carried out using a centralised controller, with all nodes communicating with each other through it.

In *peer-to-peer communication* two nodes communicate with each other, being able to send and receive data. And in a *broadcast* the centralised controller sends information to all nodes.

- **Physical Layer**: Provides the physical specification of the device and its relationship to the medium. Figure 3 shows a generic block diagram of a general physical layer. The information in binary is sent to the modulation block which will modulate it in frequency or amplitude and the resulting signal will be transmitted by the transmitter block via an LED. The channel block represents light being emitted and received by a photo-diode. The receiver will receive the modulated signal and send it to the demodulator that will separate the information from the carrier frequency, thus obtaining the coded information.

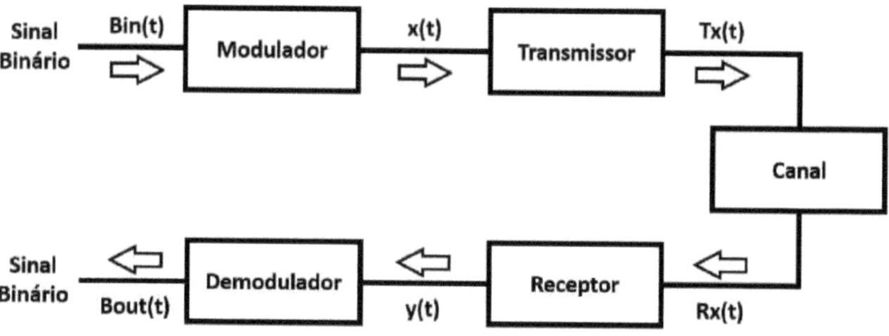

Figure 3 - Block diagram of a typical VLC system

Source: Truzzi (2016)

2.2.3 State of the Art of VLC

VLC technology has been studied and applied in different situations. Localising objects in a certain area is a well-known problem. According to Pohlmann (2010), one of the main applications of VLC, used especially in medicine, is to estimate the location of something. In the context of VLC in streetlights and traffic lights in smart cities, some works by other authors are presented below.

2.2.3.1 VLC to Estimate Location of People

Liu and Maeda (2008) proposed a scenario for visually impaired people, using VLC on indoor light bulbs to guide them through corridors. Each lamp transmits an ID via VLC which is received by a receiver worn by the person and its localisation is done by computing the distance to each lamp and estimating the person's position based on the computed distances. Figure 4 illustrates the proposed system.

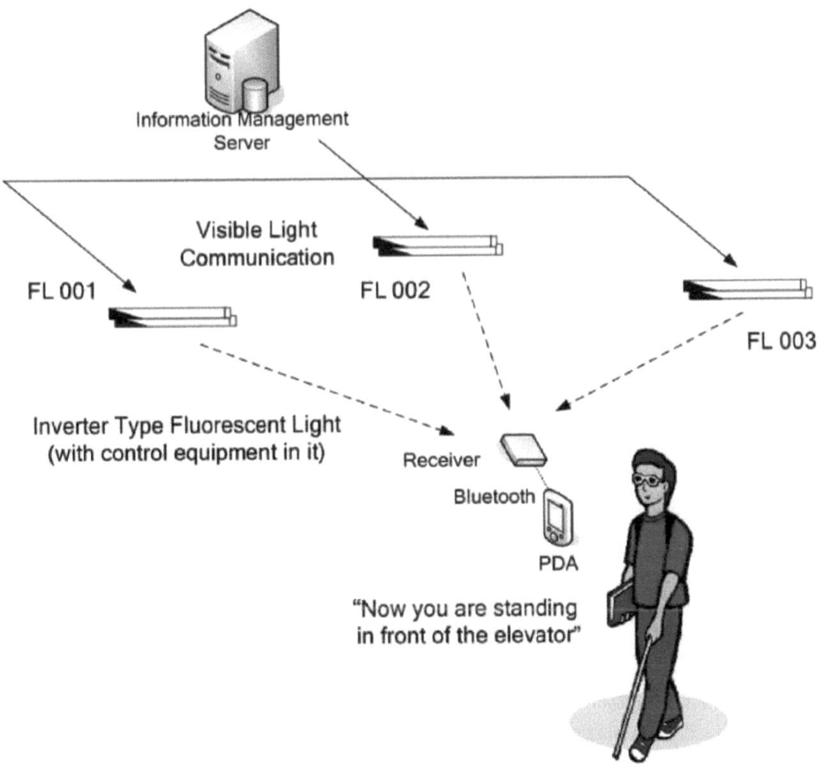

Figure 4 - VLC system for estimating people's location

Source: Liu and Maeda (2008)

The difference from this to the work proposed here is that IDs will be used on poles in open locations and each pole will be associated with a precise coordinate defined by the ID requiring no position calculation.

2.2.3.2 VLC on Smart Traffic Lights

In the work done by Kumar et al. (2012) on intelligent transport systems (ITS), VLC was used considering an LED traffic light, where the integration of the traffic light unit within the existing ITS architecture was realised. The VLC technology served to transmit the traffic light status to the vehicles over a long distance. In the results, a data communication range of more than 40m was obtained, even though the sunlight was very intense, says Kumar et al. (2012). Figure 5 illustrates the proposed system.

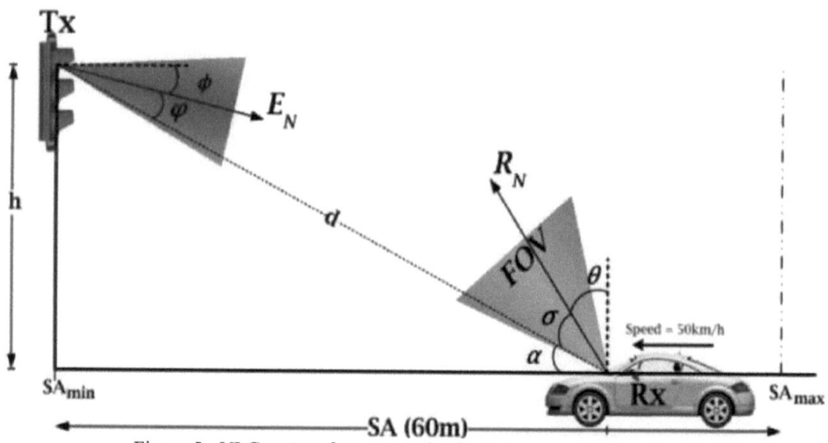

Figure 5 - VLC system for communication between vehicle and traffic light

Source: Kumar et al. (2012)

In another work, Kumar, Nero and Aguiar (2007) used VLC in advanced driver assistance systems (ADAS), which consisted of an outdoor application using VLC in LED traffic lights.

The difference in the present work is that the VLC will be applied to both traffic lights and lampposts in smart cities, all integrated into a single system.

Also in smart traffic lights, Premachandra et al. (2010) proposes a communication system between LED traffic lights and vehicles, where the traffic light transmits its state through VLC which is received by a high-speed camera in the vehicle. The luminance of the LEDs is captured in consecutive frames seeking a fast and effective communication. The in-vehicle system must first find the transmitter and then capture consecutive images to analyse the transmitted message. New techniques have been proposed to improve the communication speed of the system.

The difference between this work and the one proposed here is that it uses a photodiode instead of a camera, which reduces the cost of the project and allows maximising the speed of data transmission, as it does not require image processing.

2.2.3.3   VLC on Smart Cities

An application carried out by Boubakri, Abdallah and Boudriga (2015) aimed to build a communication architecture integrating VLC technology with smart city infrastructure. In the work, VLC was used in communication in various locations and between different devices, such as poles, traffic lights, mobile devices, vehicles and other points with lights and sensors. A network architecture structured in three layers was proposed, the first being based on VLC to detect information in specific locations in the city, the second providing communications with VLC LEDs, which consist of LEDs that send and receive data by light, and the third for communication between different points using the free optical space (FSO). Figure 6 illustrates the proposed system.

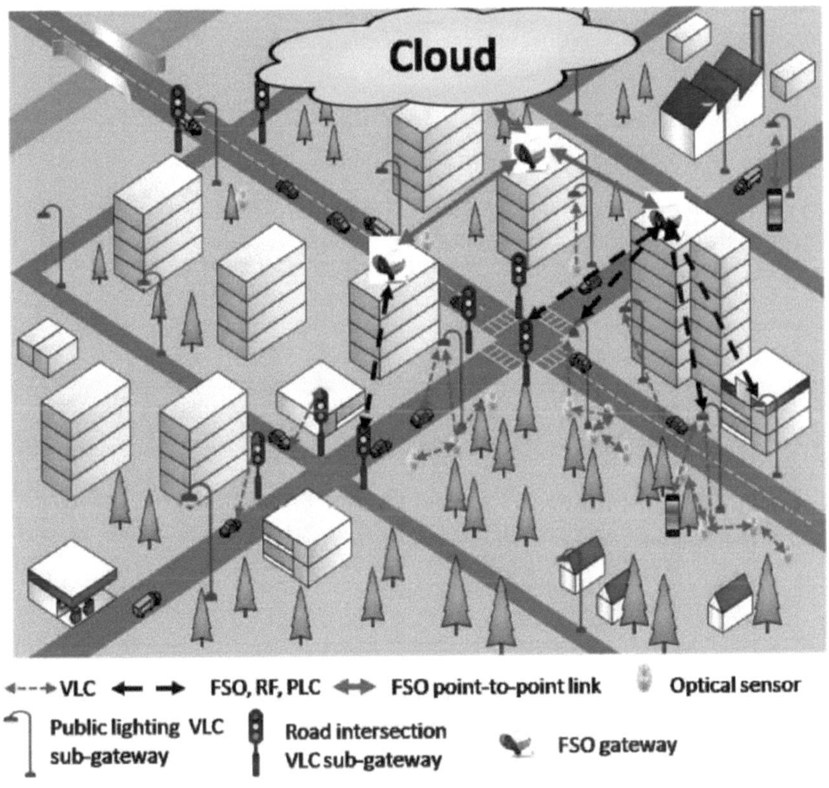

Figure 6 - Network using VLC system in smart cities

Source: Boubakri, Abdallah and Boudriga (2015).

The difference in the application presented in this article is that it seeks a simpler and more economical way to apply VLC to smart city streetlights and traffic lights, not requiring the installation of many new features.

2.2.3.4 VLC Miscellaneous System Applications

Other diverse applications of VLC are presented by Khan (2017) in his research, highlighting the following:

- **Li-Fi**: Li-Fi is the internet via light, and is a high speed bi-directional fully connected, analogous to Wi-Fi, but using visible light as the data transmission medium (VLC). Wi-Fi signals have the problem of interference with other RF signals, as well as interference with signals from aircraft navigation equipment. Therefore, in areas that are sensitive to electromagnetic radiation (such as aircraft), Li-Fi may be a better solution.

- **Vehicle to vehicle** communication: VLC can be used for communication between vehicles in traffic, and also between the vehicle and an infrastructure such as smart traffic lights. An example is a communication that prevents collision between vehicles in traffic.

- **Underwater communications**: As RF waves do not travel well in seawater because of its good conductivity, VLC communication becomes an excellent option for this situation.

- **Hospitals**: In hospitals, there are areas sensitive to electromagnetic waves, and in these areas preference has been given to the use of VLC systems so that they do not interfere with radio waves from other machines.

- **Information on signs**: At airports, bus stops and other places, you can use VLC to broadcast messages to vehicles and other mobile devices.

- **Identification system**: Visible light can be used as an ID system in different places like buildings and subways. But it can also be used in open places during the night or even during the day with low intensity lamps.

2.3 Microcontrollers

2.3.1 Origin of Microcontrollers

According to Penido and Trindade (2013), a microcontroller is a computer on a single chip. This chip contains the processor, the arithmetic logic unit, memories, input and output peripherals, timers, serial communication devices, and other items. Microcontrollers are an evolution of digital circuits as a consequence of their increased complexity. Using microcontrollers instead of logic gates is much cheaper, simpler and more compact, especially for more complex applications.

The first microcontroller, launched by Intel in 1977, was the 8048, which then gave rise to the 8051. It was programmed in *Assembly* language and has a powerful set of instructions. It has internally a CPU (Central Processing Unit, a PROM memory (Programmable Read Only Memory), a RAM memory (Random Access Memory), a set of I/O (Input and Output) lines, and a set of auxiliary devices for operation, such as clock generator, counters, timers, etc (PENIDO; TRINDADE, 2013). In general, the other microcontrollers follow a similar structure.

There are important differences between a microcontroller and a microprocessor. For a microprocessor to work, other components are needed to receive and send data, and the microprocessor is like the heart of the computer. A microcontroller, on the other hand, has all the devices necessary for its operation in a single integrated circuit. Thus, other external components are not needed in the applications, because the microcontroller already has all the necessary peripherals, which saves time and space in the construction of devices (ANTONIO, 2006).

Microcontrollers now replace integrated circuit assemblies in most applications. Microcontrollers are used in automated products and devices, implantable medical devices, remote controls, office machines, home appliances, power tools, toys and other embedded systems.

2.3.2   PIC Microcontrollers

PIC (*Peripherical Interface Controller*) microcontrollers are a well-known family of microcontrollers, manufactured by the company Microchip. Microchip is a major worldwide manufacturer of microcontrollers, and is responsible for producing PICs. These micro-controllers have RISC (*Reduced Instruction Set Computer*) technology and processors with reduced instruction sets. In general, there are 35 simple instructions that execute in one or two machine cycles. There are PICs of 14, 16 and 32 bits, of 8, 14, 16, 28 and 40 pins in the encapsulation, which allows a wide range of applications (PENIDO; TRINDADE, 2013).

The acronym PIC is the name that Microchip adopted for this family, which stands for Peripheral Interface Controller. According to Antonio (2006), the PIC has internally all the typical devices of a microprocessor system, being:

- **Central Processing Unit (CPU)**: Its purpose is to interpret and process programme instructions.

- **Programmable Read Only Memory (PROM)**. Used to permanently store the program instructions written by the programmer.

- **Random Access Memory (RAM)**: Used to store the temporary variables used by the program during its execution.

- **Input and Output Lines (I/O)**: Which serve to control external devices or receive pulses from sensors, switches, among others.

- **Auxiliary Devices**: A number of other devices in a very small space, such as clock generator, timers, etc.

With the presence of these devices in a small space, the designer has a large working area and many advantages in using the microprocessor system, where in a short time and with few external components it is possible to do what would be very labour intensive using traditional components.

There are several models of PICs, each of which can adapt to the requirements of specific projects, differing by the number of inputs and outputs, number of pins, and other contents of the device (ANTONIO, 2006). The family starts with a small model identified by the acronym PIC12Cxx with 8 pins, until reaching larger models with the acronym PIC17Cxx with 40 pins. On the Microchip website[1] , detailed descriptions of the PIC typology can be found, with extensive and varied technical information, support *software*, application examples and available updates.

Figure 7 shows the internal circuit block diagram of one of the simplest PICs available, the PIC16F84A. This diagram shows all the basic elements of one of the first PIC microcontrollers that were built, serving as a basic model for all other PICs, the difference being that they will have more memory and more internal devices than those seen in the figure.

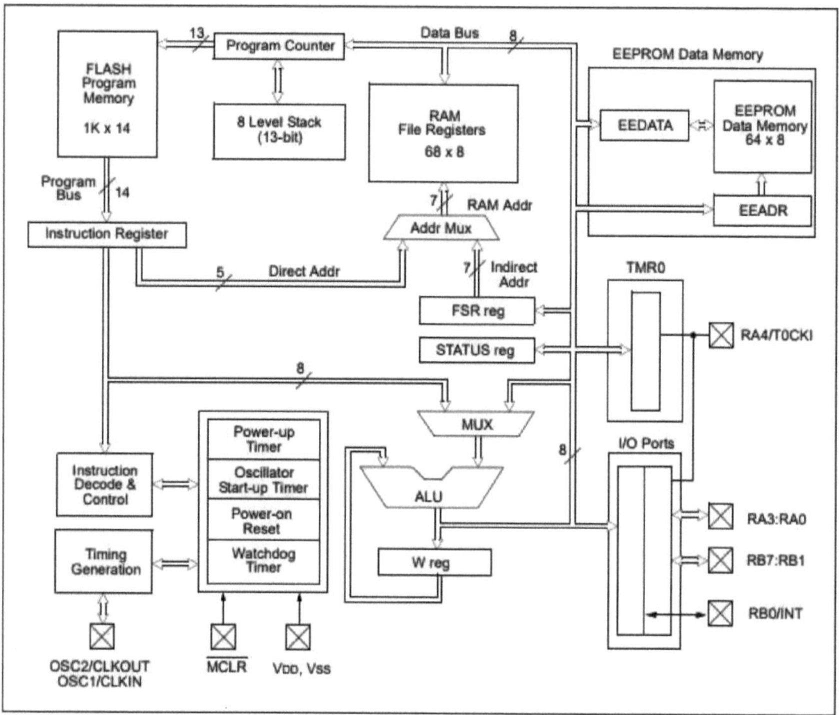

Figure 7 - PIC16F84A block diagram

Source: PIC16F84A Datasheet

2.3.3 Programming PICs in C

The PIC is a programmable device, and the purpose of the programme is to leave instructions so that the PIC can do activities defined by the programmer. A programme is composed of a set of sequential instructions, where each one identifies a specific function that the PIC will perform. The programme can be written in Assembly or C language, through any computer using any word processor that can generate files in ASCII standard. This file will be called *source* or assembly code (ANTONIO, 2006).

The code, once written, will have its mnemonic instructions and all other conventional forms of writing converted into a series of numbers (the opcode) recognisable directly by the PIC. This process is performed by a compiler (ANTONIO, 2006). In this work, the language used was C and the compiler was CCS[2] .

Figure 8 shows a flowchart of compiling a C or Assembly programme.

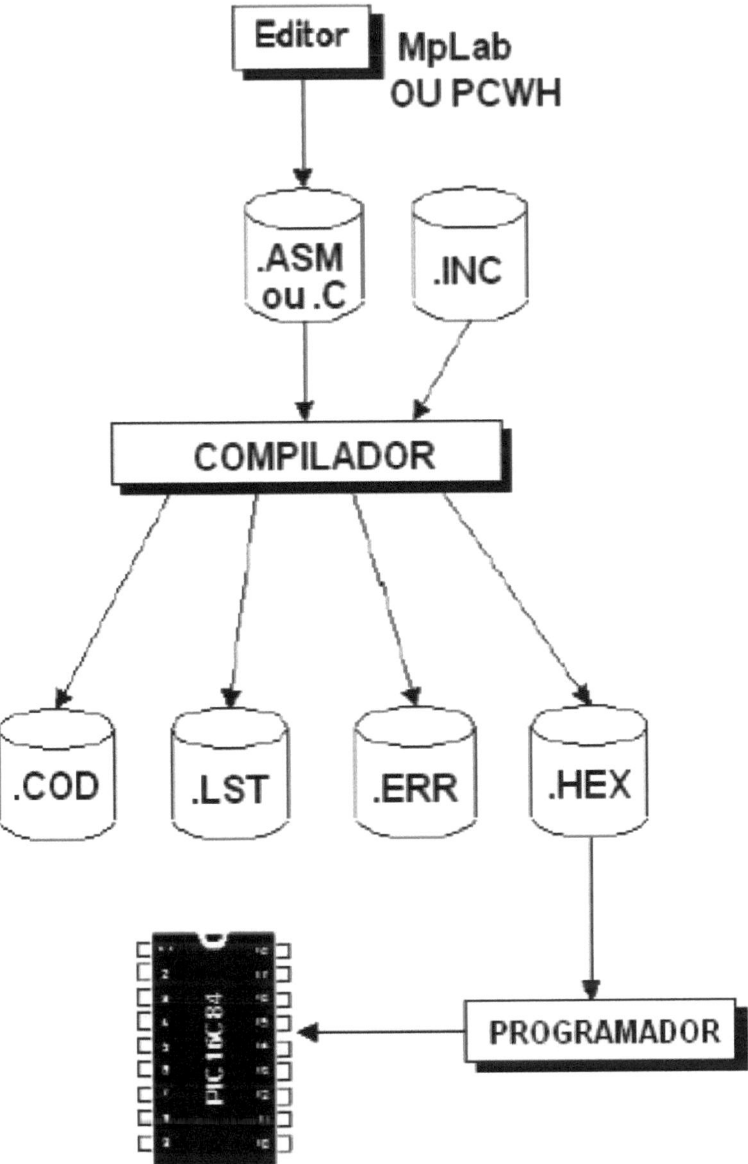

Figure 8 - Flowchart for compiling a C or Assembly programme
Source: Antonio (2006)

The editor is the programme responsible for allowing the user to write his programme in the desired language, generating an .ASM or C (Assembly or C) file, and an .INC file, showing the files included in the

17

project. These files will pass through the compiler which will generate the .HEX file containing the hexadecimal code to be written in the microcontroller, and auxiliary files. Then the programmer, which consists of an electronic module for recording PICs, will write the binary values corresponding to the .HEX file into the microcontroller's programme memory, which can finally be placed in a circuit and execute the programme developed by the user.

The C language was created in 1972 by Dennis Ritchie, from Bell Laboratories, consisting of an intermediate level language between the Assembly language and high-level languages (PEREIRA, 2003). Pereira states that until they developed the C language, there were no high-level languages capable of creating operating systems and other low-level *software*, which led developers to use Assembly. It was from the needs of rewriting the UNIX operating system that the C language emerged.

The use of the C language to programme PICs is a natural choice, says Pereira (2003). Most microcontrollers available on the market have C language compilers for software development. The use of C allows building much more complex applications than in Assembly. In addition, the use of the C language allows a great speed in the creation of new projects, due to the programming facilities offered by the language and its portability, allowing to adapt programmes from one system to another without much effort (PEREIRA, 2003).

2.4   Communication Systems

2.4.1   Serial Communication

Messages transmitted in a serial communication are broken into smaller parts and transmitted sequentially bit by bit. Bit-serial transmission converts the message into one bit at a time through a channel. Each bit represents a part of the message. The individual bits, upon arrival at the destination, are rearranged to make up the original message. A channel generally allows only one bit to pass at a time. Bit-serial transmission, also called serial transmission, is the method chosen by many computer peripherals (CANZIAN, 2013).

There is also byte-serial transmission, which converts 8 bits at a time through 8 parallel channels. This type of transmission is 8 times faster than bit-serial transmission but requires 8 channels, making the cost 8 times higher to transmit the same message.

Canzian (2013) states that serialised data is not sent uniformly over a single channel, regular information packets sent followed by a pause are used. Binary data packets are sent with varying pause lengths from one packet to another, until the message has been fully transmitted. The receiver must know the appropriate time to read the individual bits of this channel, know exactly when a packet starts and how much time elapses between bits. How timing occurs depends on the type of communication. According to Rocha and Gameiro (2015), communication systems can operate in a synchronous or asynchronous transmission mode.

2.4.2   Synchronous Communication

In synchronous communication mode, all transmitted bits are information bits and therefore the transmission efficiency is 100%. The clocks of the sender and receiver are phase and frequency dependent.

There are mechanisms that locally control the frequency and phase of the local clock, to follow the transmitted clock intrinsically in the data. This mechanism is called clock synchronisation (ROCHA; GAMEIRO, 2015).

Thus, in synchronous communication, the receiver is said to be synchronised with the transmitter since the timing of both is known and is the same. Accurate data transfer is guaranteed. Synchronism failures will cause data corruption or loss (CANZIAN, 2013).

In synchronous systems, a clock can be used synchronising the transmitter and receiver, so that with each clock pulse the transmitter sends a bit and the receiver receives the respective bit based on the clock. For this, in the synchronous system two transmission channels are required, one for the data and one for the clock.

Figure 9 shows an example of a basic circuit with synchronous data sender and receiver.

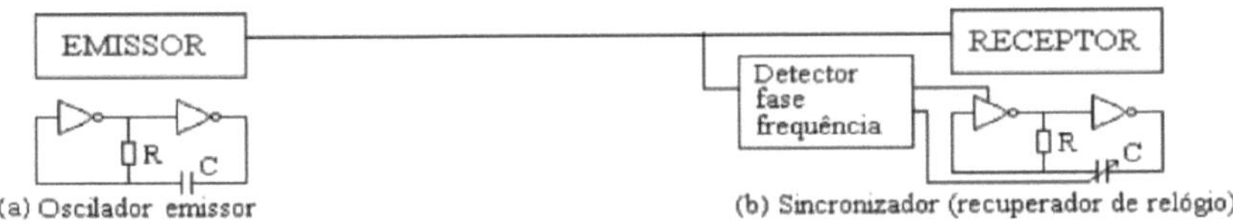

Figure 9 - Basic circuit of synchronous transmitter and receiver

Source: Rocha and Gameiro (2015)

2.4.3 Asynchronous Communication

In asynchronous communication, the length of the data packet must be very short to minimise the risk of the transmitter and receiver oscillators varying. When crystal oscillators are used, synchronisation can be guaranteed over the 11 period bits. With each new packet sent, the *start* bit restarts the synchronisation, which means that the pause between packets can be long. The transmitter and receiver oscillators are as identical as possible. The receiver oscillator has an input to reset its phase when the *stop* bit occurs and correctly collects the *start bit*. After that, it oscillates freely in the bits of the information block (ROCHA; GAMEIRO, 2015).

In asynchronous systems, information travels over a single channel. The transmitter and receiver must be set up in advance for communication to take place correctly. The oscillators must be very precise and identical in the transmitter and receiver. In most common serial protocols, data is sent in small packets of 10 or 11 bits, of which only 8 constitute the message. The data packet always starts with logic level 0 (*start* bit) to signal to the receiver that transmission has started. The *start bit* initialises an internal timer in the receiver warning that transmission has started and that clocking pulses will be required. After this, the 8 message bits are read at the specified baud rate. At the end of the packet, the *stop bit* indicates its termination (CANZIAN, 2013).

Figure 10 shows an example of a basic circuit with asynchronous data sender and receiver.

Figure 10 - Basic circuit of asynchronous transmitter and receiver

Source: Rocha and Gameiro (2015)

2.5 Global Positioning Systems (GPS)

2.5.1 GPS Definition and Operation

The *Global Positioning* System (GPS), also known as NAVSTAR-GPS (*Navigation Satellite with Time And Ranging*), is a radio navigation system developed by the United States Department of Defence, whose initial objective was to be the main navigation system of the American army (CARVALHO; ARAUJO, 2009).

According to Carvalho and Araujo (2009), GPS is programmed to provide two- or three-dimensional coordinates of points on the ground, as well as the speed and direction with which we move between these points. GPS is intended to assist in navigation activities and in carrying out geodetic and topographic surveys. It operates 24 hours a day, in any weather situation. However, certain climates can cause interference with the quality of GPS data. The system is programmed so that at least 4 satellites can be observed at any time of the day and anywhere on the planet, ensuring position determination at all times, anywhere on the planet (CARVALHO; ARAUJO, 2009).

The GPS system is based on determining the distance between a point called the receiver and the satellite reference points. Once the distance between the receiver and three reference points is known, the relative position of the receiver can be determined by intersecting three circles whose radii are the distances measured between the receiver and the satellites. The entire GPS system is based on the triangulation of satellites, whose signals can be received by means of receivers of various types. Each satellite emits a signal in code that identifies it. The receiver interprets the received signal and calculates its distance to the sending satellite. The system is able to calculate the position in two dimensions (latitude and longitude), or also in three dimensions (latitude, longitude and altitude). However, for a perfect calculation, there must be no errors in the time measurements made.

GPS needs at least two properly functioning satellites to calculate the two-dimensional position of the receiver. Atomic clocks on the satellites are needed for measurement. The biggest disadvantage of GPS is the lack of signal in specific locations, such as car parks and other enclosed places.

One of the benefits of the developed project, therefore, is to complement the GPS in specific locations, so that in case the GPS fails in these locations, the pole ID system will serve as a reference for the vehicle to know its location, without the need for GPS presence in the city.

2.6 BER and PER Error Rates

2.6.1 Bit Error Rate (BER)

The bit error rate (BER) is a parameter used in digital data transmission systems. BER is widely used in radio, fibre optic, ethernet, and other systems that transmit data over a network containing various types of noise. There are many differences in how these systems work and how BER is affected, but the basis for calculating the rate is the same (POOLE, 2015).

BER, unlike many other forms of evaluation, assesses the complete performance of the system, including the transmitter, receiver and transmission channel. In this way, BER allows the actual performance of the system to be tested, rather than testing the components in isolation. According to Poole (2015), the general calculation of BER is defined by Equation 2.1.

$$BER = \frac{Numero\_de\_bits\_errados}{Numero\_de\_bits\_transmitidos} \tag{2.1}$$

BER is very useful to validate VLC systems, being used to validate the whole system in several references found. Kumar et al. (2012) used BER to validate the VLC system of communication between traffic lights and vehicles, and presents the rate graph as an important result of the developed project.

2.6.2 Packet Error Rate (PER)

The Packet Error Rate (PER) is a parameter also used in digital data transmission systems to test the percentage of packets that were received with errors. The general PER calculation is defined by Equation 2.2.

$$PER = \frac{Numero\_de\_pacotes\_errados}{Numero\_de\_pacotes\_transmitidos} \tag{2.2}$$

The PER rate was also used by Kumar et al. (2012) to validate the communication between traffic lights and vehicles, presented as an important result of the developed project.

CHAPTER 3

Project Developed

In this chapter, the development of the proposed solution is presented.

3.1 Context Description

For the project, a city whose streetlights and traffic lights have LED lamps was considered as a context. The distances between the poles are sufficient so that the light of one pole does not interfere with the light of the other pole or the traffic light, and to ensure this, a threshold of light intensity (defined by fine-tuning the sensitivity of the receiver) was considered, so that at the points of intersection of the lights between one pole and another, there is no reception of the messages transmitted by the lights, since if this were done, errors would occur due to the interference of one pole in the other.

For the implementation of the prototype, a small section of the city was considered, consisting of a road with four poles and a traffic light, whose proposal consists of using only one lamp of the traffic light to transmit its state to the smart vehicles, but also being able to reuse the lamps of the existing traffic lights. Both the lampposts and the traffic lights must have sufficient intensity to be detected by the receiver in any horizontal position on the road. Figure 11 illustrates the section of the city considered.

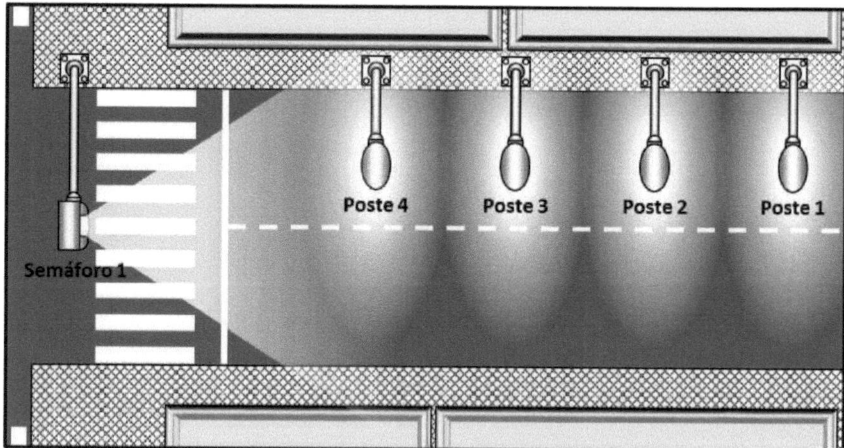

Figure 11 - Section of the city considered as context

Source: Author

3.2   System Description

The system consists of a transmitter circuit in the infrastructure, connected to all the poles and traffic lights in the city, and a receiver circuit in each vehicle, to receive messages sent by the city's poles and traffic lights.

3.2.1   Infrastructure

In the city's infrastructure, considering the context presented, it is proposed that each pole constantly sends, through VLC technology, an ID (identifier number) that represents its position in the smart city. This ID, in the prototype presented, consists of a decimal integer that is transmitted through a sequence of 8 bits in the light of the pole, at a transmission speed sufficient so that the variation in light intensity is not perceived by the naked eye (raw data are presented in the results). As there are four poles in the example stretch, each pole sends its respective integer, which in this case ranges from 1 to 4.

For the traffic light, it must constantly send an integer indicating its status (closed or open), also through VLC technology, by means of a single lamp that is constantly switched on. In the example section, two numbers not used by the poles were used, 8 for open traffic lights and 9 for closed traffic lights. The traffic light automatically changes its value from time to time, alternating between open and closed. A time of 20 seconds was used for the example section.

The sequencing and synchronisation of the sending of data by the light is done for all poles and traffic lights by a single central that has a microcontroller connected to each light of each pole and traffic light, so that this central generates all the IDs for all the poles and also the numbers for the traffic light.

3.2.2   Intelligent Vehicle

In the intelligent vehicle, a receiver circuit is proposed whose processor has an oscillator that does not need to have the same clock as the transmitter oscillator, but must be configured for the same transmission rate as the transmitter, which in this case uses the PIC UART module. The communication will be done through a single channel which is the light.

When the vehicle passes through the area of action of the light source, the receiver circuit will receive the pole ID through the VLC, and will know, based on the pole ID, its location in the city, since each pole has a unique ID. In the implemented prototype, the received byte is shown in a sequence of 8 LEDs and the pole ID is shown in an LCD display, in the receiver circuit. When the vehicle approaches the traffic light, it will receive in an analogue way, the information sent by the traffic light, of open or closed. It is up to the vehicle system to decide its action regarding the status of the traffic light and even the pole IDs.

3.2.3   System Illustration

Figure 12 illustrates the operation of the system in the city. The transmitter circuit is connected to all poles and the traffic light and each vehicle has an integrated receiver circuit. Above each vehicle there is a phototransistor and its field of view (FOV) is displayed in the figure.

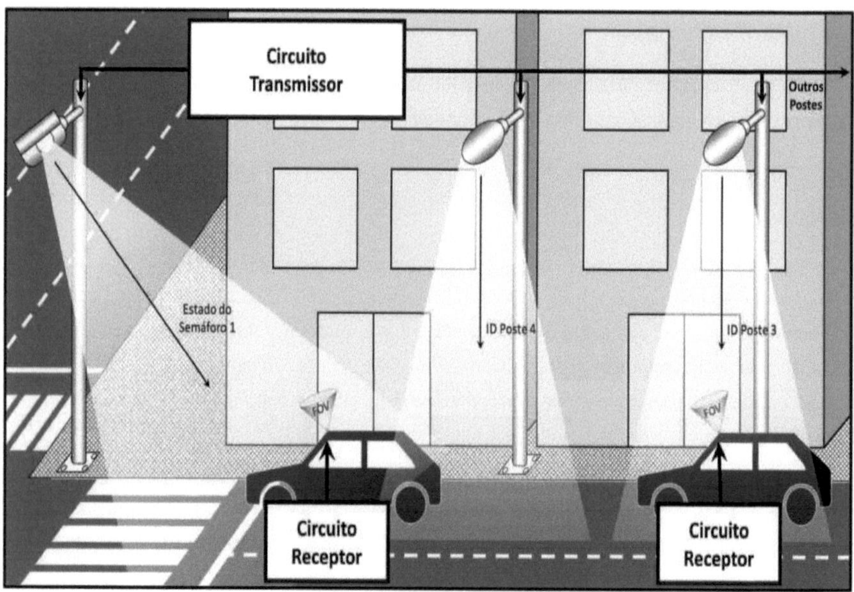

Figure 12 - Illustration of the system

Source: Author

3.3 Implemented Electronic Circuits

In this section, the operation of the implemented transmitter and receiver circuits is described. The lists of materials used for these are presented in Appendix A.

3.3.1 Transmitter Electronic Circuit

For the transmitter circuit, a PIC16F870 microcontroller with a 4.43MHz crystal and auxiliary components was used. Five high-brightness LEDs were used, of which four (connected to pins LP1+/LP1- to LP4+/LP4-) represent the four posts and the fifth (connected to pins LP5+/LP5-) represents the traffic light. For each LED, a BC548 transistor and biasing components were used in order to guarantee a power in the LEDs brightness sufficient to reach the desired communication distance (raw data in the results), without damaging the LED or the microcontroller output. The microcontroller was programmed in C language using the MPLAB development environment and the CCS compiler. Each LED therefore constantly sending a byte informing its ID. To power the circuit, a module with two pins (V1+/V1-) was implemented to connect a battery or external DC power supply, two pins to connect an on-off switch (S1A/S1B), a LM7805 voltage regulator to fix the circuit voltage at 5Vdc, which is the working voltage of the PIC, and capacitors to filter noise from the supply voltage (CF1,CF2 and CF3). A green diffuse LED (DG1) with its respective drive resistor (RDG1) was also added to signal that the circuit is sending the IDs. When the circuit is working correctly, this LED is blinking.

24

The high-brightness LEDs used support voltages around 3V and currents around 30mA. The circuit is designed for a voltage of 3V and current of 35mA. The calculation of the collector resistor is done by RCPi=(5V-3V)/35mA=57ohm (Commercial value: 56ohm). The calculation of the base resistor of the transistor, with Beta=110 and VBE=0.77V, was made by RBPi=(5V- 0.77V)/(35mA/110)=13.2Kohm (BOYLESTAD; NASHELSKY, 2004). In the implemented prototype, 8.2Kohm base resistors were used (slightly below the calculated value). The LED signalling resistor RDG1 was calculated to maintain a voltage of 1.5V and current of 20mA in the LED, by the equation RDG1=(5V-1.5V)/20mA=175ohm (Commercial value: 180ohm).

Using the same drive components, the luminescence of each LED was measured for different LED colours, using the Ourolux Luxmeter Android app[1] . Although it is not a robust luxmeter, as it is only a comparison between luminescences, the app met the needs of the project. Table 1 shows the luminescence values of each colour, at night, at a distance of 1m between the LED and the mobile phone sensor. Although the blue LED has a higher luminescence, white LEDs were used because it is the colour used in streetlights in cities. Using a focusing lens on the white LED, the luminescence at this same distance increased to 577Lx.

Table 1 - Luminescence of LED colours

LED colour	Luminescence in Lx
Yellow	38Lx
Blue	77Lx
White	47Lx

Source: Author.

Figure 13 shows the schematic diagram of the transmitter circuit. Figure 14 shows the project of the board designed in Proteus ARES. And Figure 15 shows the 3D visualisation of this board.

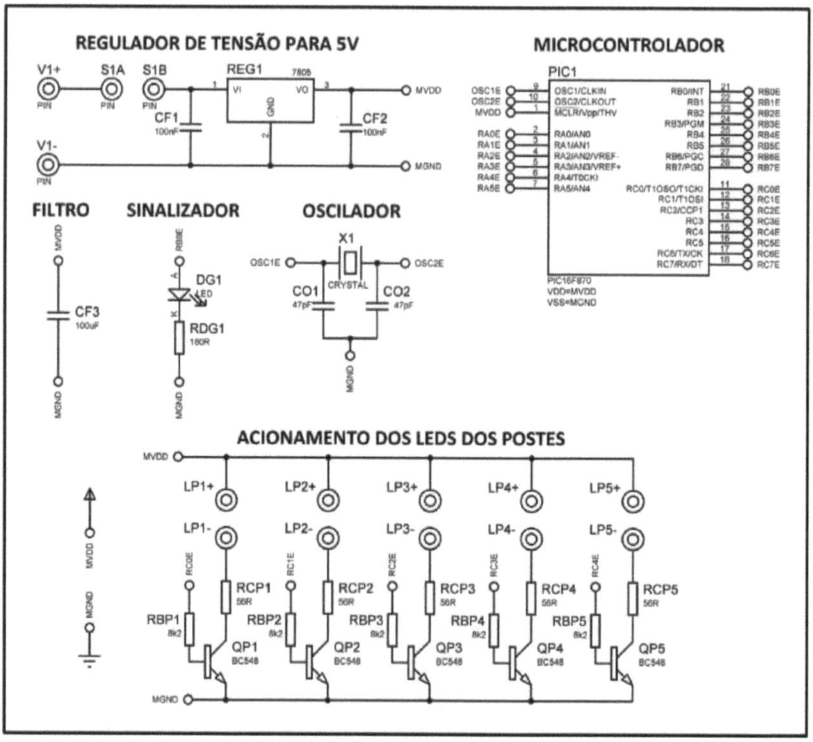

Figure 13 - Transmitter circuit (Infrastructure)

Source: Author

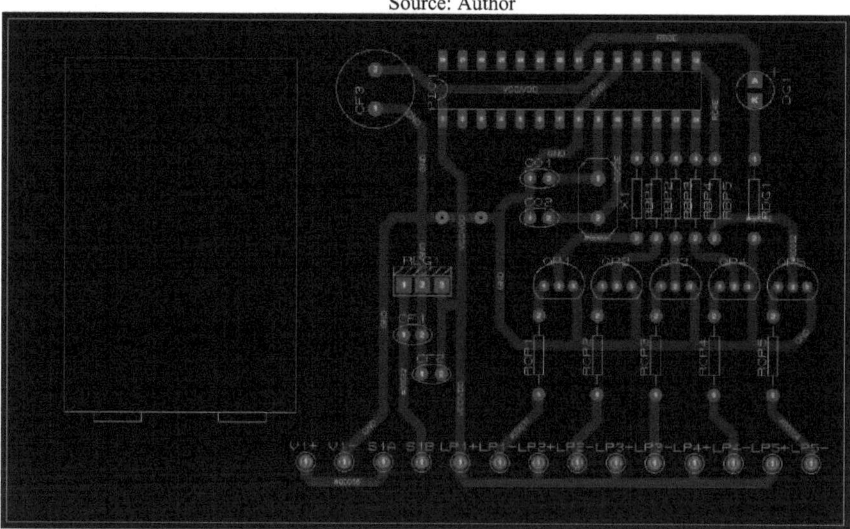

Figure 14 - Transmitter circuit board design

Source: Author

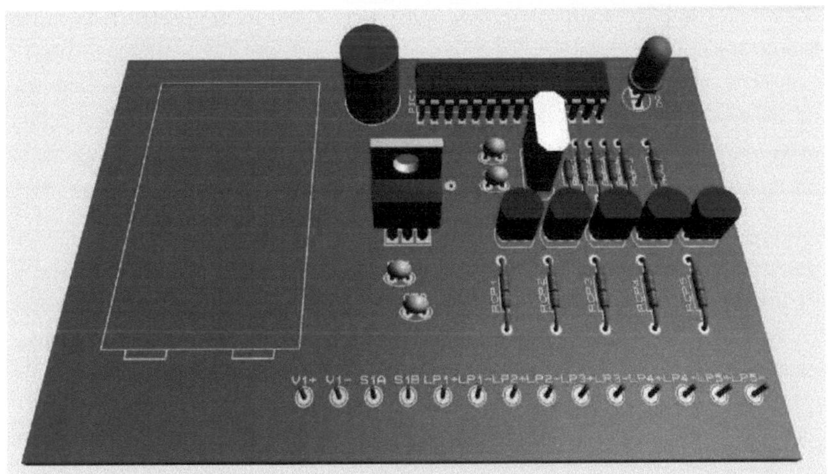

Figure 15 - 3D visualisation of the transmitter circuit board design
Source: Author

As can be seen, all the pins of the circuit are positioned at the bottom of the
board, to connect all the external components, which are the battery, the main switch and the 5 high-brightness LEDs, via terminals. A space has been reserved on the board to connect a 9V battery.

3.3.2   Receiver Electronic Circuitry

For the receiver circuit, a PIC16F870 microcontroller with a 4.43MHz crystal and auxiliary components was also used. For light reception, a 5mm transparent receiver phototransistor (FTC/FTE pins) was used. The emitter terminal of this phototransistor (FTE) is connected to a 4.7Kohm resistor (R5) in series with a 100Kohm trimpot (RF1+/RF1M/RF1-) whose other terminal is connected to the negative of the source, allowing a sensitivity adjustment for the phototransistor. The common point of these is connected to the non-inverting input of an operational amplifier (LM358 IC), configured with a voltage gain Av=23, through the 1Kohm and 22Kohm resistors (R1 and R2), calculated by Av=(1+22Kohm/1Kohm)=23 (BOYLESTAD; NASHELSKY, 2004). At the output of the amplifier there is a circuit with a 180ohm resistor (RDG1) and a green diffused LED (DG1) that serves to signal that the light received by the photo diode was sufficient to generate logic level 1 at the output of the amplifier. This output is then connected to a digital input of the microcontroller (Pin RA0R), which is configured to receive serial information. The calculation of triggering all LEDs was done in a similar way to the transmitter's signalling LED.

No information was found regarding the opening angle of the phototransistor used. To know the approximate angle, the maximum angle of inclination of the phototransistor with respect to the LED was measured, until the phototransistor was fully open, at a distance of 1m between the phototransistor and the LED. The maximum angle obtained was approximately 28 degrees.

Eight diffuse LEDs (DB1 to DB8) were used to display the received byte in binary, and an LCD display to display in text form, the ID of each pole and the status of the traffic light, with previously defined messages associated with each received value. For programming the microcontroller, the same resources used for the transmitter were used. The information received by the phototransistor is displayed both on the LEDs, in binary form, and on the LCD display with the respective message associated with each received data. To power the circuit, a voltage regulator module identical to the one implemented in the transmitter was implemented. An on/off switch with lock was also added to perform tests on the circuit, especially to collect the received bits and packets and calculate the BER and PER error rates. The function of the switch was to start counting the bits and packets (two different situations).

Figure 16 shows the receiver circuit diagram. Figure 17 shows the project of the board designed in Proteus ARES. And Figure 18 shows the 3D visualisation of this board. As can be seen, all the pins of the circuit are positioned on the right side of the board, to connect, through terminals, all the external components, which are the battery, the main switch, the phototransistor and outputs to send the data to a Raspberry.

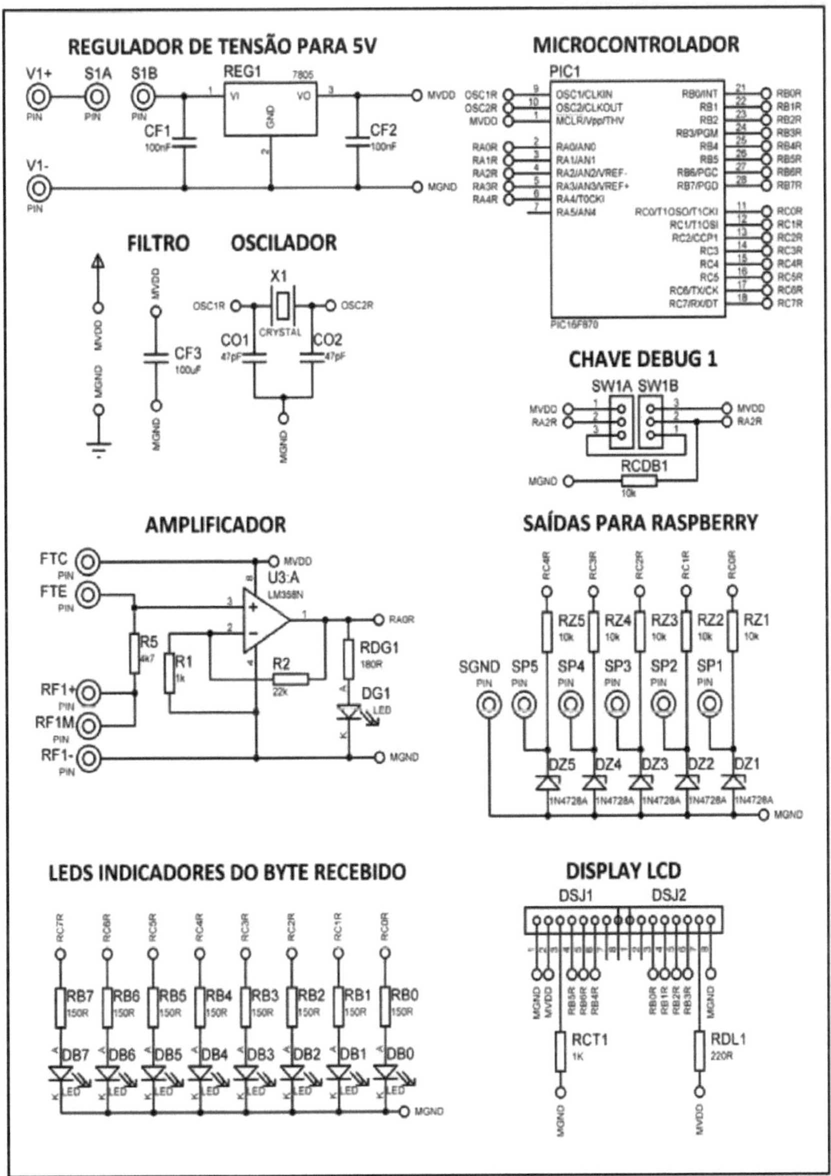

Figure 16 - Receiving circuit (Vehicle)

Source: Author

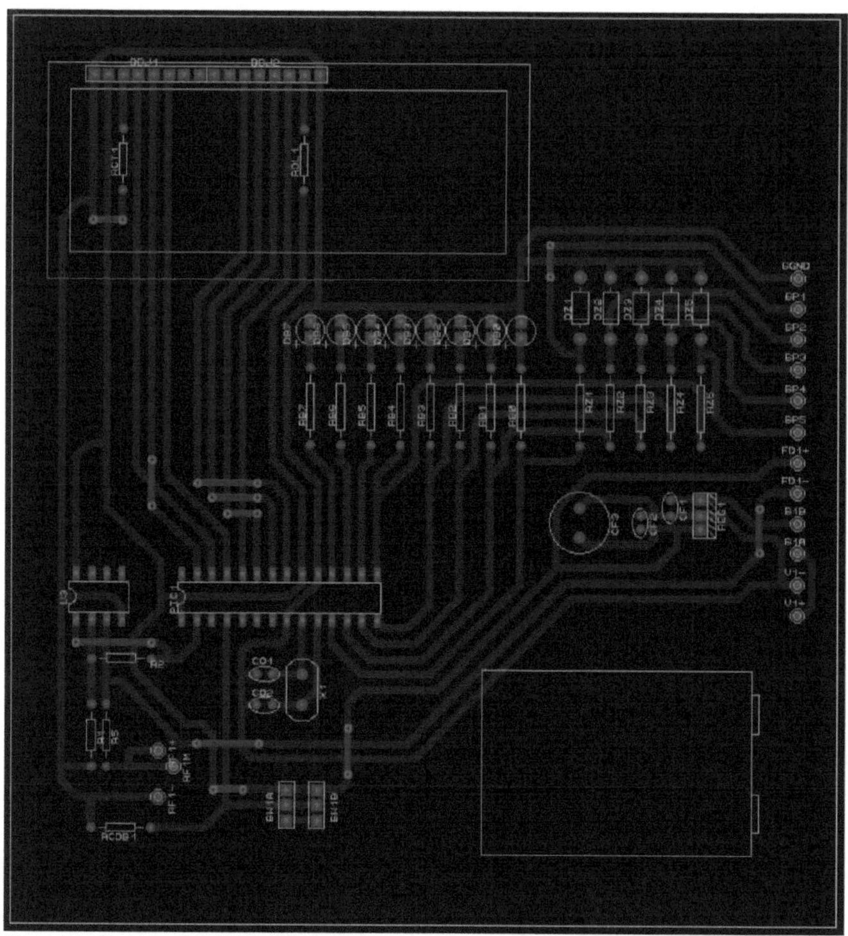

Figure 17 - Receiver circuit board design

Source: Author

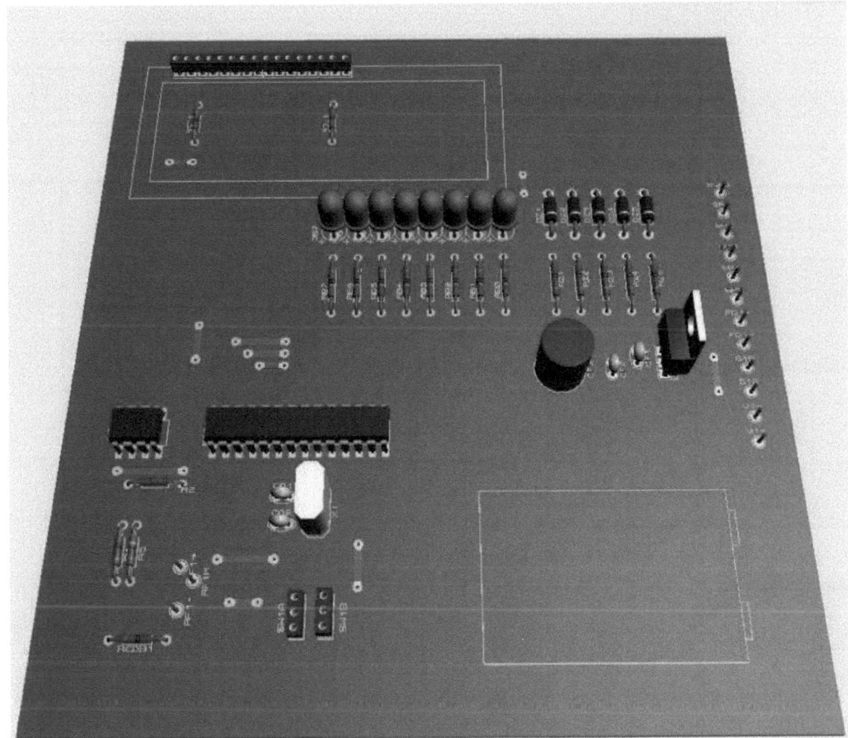

Figure 18 - 3D visualisation of the receiver circuit board design

Source: Author

Additionally, 5 3.3V outputs are added to send information from each pole and traffic light to the Raspberry. Outputs SP1 to SP4 determine which pole the vehicle is on (the respective output will be at level 1), and output SP5 sends a high logic level if the vehicle is in front of the closed traffic light. For this, we used 1N4728A zenner diodes of 3.3V (input voltage of the Raspberry used) and 10Kohm resistors in series with them (which serve to carry the complementary voltage to the zenner voltage). This section of the circuit was designed in addition to the objectives of the present project so that such transmitted information serves for the evolution of this work in the context of its inclusion in an intelligent vehicle.

3.   4Microcontroller codes

Appendix B presents the codes implemented in the transmitter and receiver microcontrollers. These were developed in C language using the CCS compiler and the MPLAB development environment. The codes were divided into blocks separated by lines, and each block is described in this section.

3.4.   1Transmitter Code

The transmitter code was divided into four blocks, these being the settings, addresses, global variables and main loop blocks. Figure 19 shows a diagram illustrating the logic between these blocks.

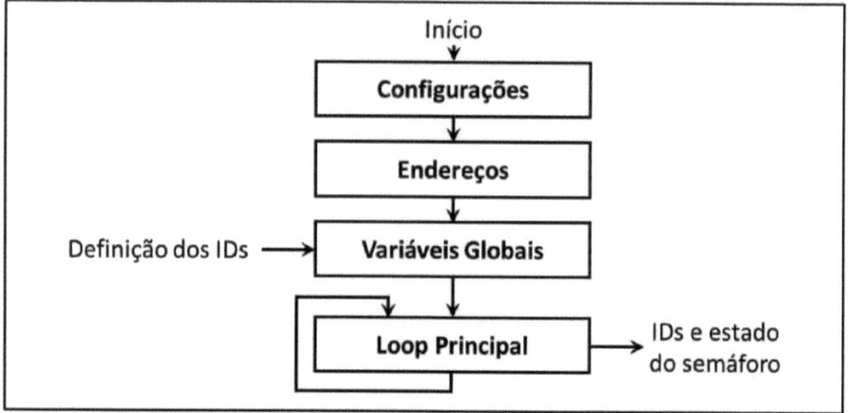

Figure 19 - Block diagram of the transmitter code

Source: Author

The first block (Settings) contains the PIC settings. The #FUSES directive configures the use of crystal oscillator (HS), disables watchdog timer (NOWDT), disables powerup timer (NOPUT), disables voltage drop detection (NOBROWNOUT), and disables low voltage programming (NOLVP). Then the use of delays (#USE delay) is configured with a clock of 4.43MHz. And then the serial outputs are configured (#USE rs232). 5 rs232 channels were used, with a frequency of 1800bps, on pins RC0 to RC4, and pin RC7 was configured as a receiver. However, as the circuit only transmits, this pin is configured (mandatory) but not used. 8bits of data were defined, with no parity bit, and the channels were assigned the names chan1 to chan5.

The second block (Addresses) contains the names assigned to the addresses of the internal registers of the PIC, these are PORTA, PORTB, PORTC, TRISA, TRISB and TRISC, and also names are assigned to specific bits, in this case only the LED1 bit which is bit 0 of PORTB.

The third block (Global Variables) contains the declared global variables. The variables clksem1i and clksem1x configure a counter from 0 to 500 used for the traffic light. The variables clkled1i and clkled1x configure a counter from 0 to 10 for the beacon LED. The variables post1 to post5 set the bytes with values that define the IDs of the posts. For bytes, the CCS compiler generally uses the char data type.

The last block (Main Loop) contains the main function of the programme. It counts the time of the traffic light and the beacon LED, and sends the bytes of each pole, each on its respective channel defined in the settings.

3.4.2   Receiver Code

The receiver code was divided into 6 blocks, these being the settings, addresses, global variables, LCD functions, initialisation and main loop blocks. Figure 19 shows a diagram illustrating the logic between these blocks.

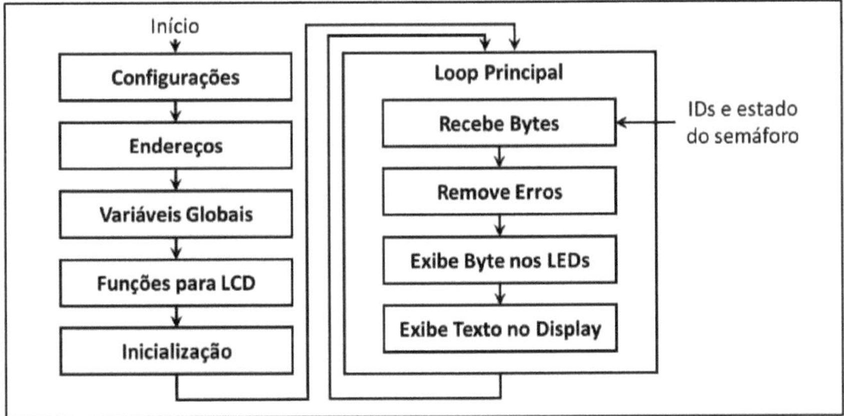

Figure 20 - Block diagram of the receiver code

Source: Author

In the first block (Settings), there are the same settings used in the transmitter, with the difference that in the receiver there is only one rs232 channel, configured at the same bit rate (1800bps) as the transmitter, with the RA1 pin as transmitter (configured mandatorily but not used), and the RA0 pin as receiver, and without parity bit.

The second block (Addresses) is similar to that of the transmitter, containing the addresses of PORTA, PORTB, PORTC, TRISA, TRISB and TRISC, the control bits of the LCD display (RS, RW, and EN) and the test key bit S1.

The third block (Global Variables) contains the global variables. These variables are just two character vectors that store the messages of the top and bottom line of the LCD display (txt1[16] and txt2[16]).

The fourth block (Functions for LCD) contains functions to control the LCD display. The inst(char v) function receives an instruction and sends it to the LCD. The function data(char v) receives a data and sends it to the LCD. The send_byte(char v, int t) function receives a byte and its type (instruction/data) and sends it to the LCD. The lcd_init() function initialises the LCD. The lcd_gotoxy(char x, char y) function receives the x and y coordinates and positions the cursor at that position on the LCD. The function lcd_putc(char c) receives a character and displays it on the LCD. The function lcd_txt(char t1[]) receives a vector of characters and displays it on the LCD. The lcd_int(long val) function receives an integer, converts it to text and displays it on the LCD. And the function lcd_txtint(char t1[], long val) receives a text and an integer, and displays them on the LCD.

The fifth block (Initialisation) contains the init() function which serves to initialise the system, configuring the PIC PORTs, the LCD and displaying the message "VLC SMARTY CITY" on it.

The sixth block (Main Loop) contains the main function of the programme, and has been subdivided into four blocks. The first one receives the byte sent by the transmitter, through the serial channel chan1. The second one checks and removes possible errors, using auxiliary variables that aim to remove incorrectly received data, so that only when the message is one of the possible predefined messages, it will be considered as a valid message. In addition, to reduce errors, the system only considers the data as valid after receiving it twice consecutively, which reduces the transmission speed but increases the accuracy of the received data. The third one displays the byte on the LEDs in binary. And the last one shows the respective message associated with the byte on the LCD display.

3.4.3 Communication Protocol

Initially, a communication protocol containing 19bits had been defined, where the first 10 were the preamble, the next the message type and the last 8 the message. However, when migrating to the use of the PIC serial communication modules, the management of the communication bits is done by the PIC system itself. Thus the final protocol consists of only receiving the integer byte containing a decimal number that can go from 0 to 255, and all the information is transmitted in a single 8bit packet, which maximises the communication speed. For a system containing more than 256 posts, one could use more communication bits or transmit the information in more than one packet, such as using two packets to transmit a single ID. However, in this prototype this was not necessary.

CHAPTER 4

Results Achieved

This chapter presents the results obtained with the development of the prototype. To better present the results, this chapter was organised in two sections. The first exposes the electronic circuit developed in different configurations (poles and traffic light) in order to demonstrate its real operation. The second shows the raw data of the maximum range and other parameters of the prototype and seeks to validate the proposed system according to the BER (Bits Error Rate) and the PER (Packet Error Rate) for different distances and with different interferences.

4.1 Qualitative Results

4.1.1 Final Physical Design

Figure 21 shows the final circuit of the transmitter centre, without the LEDs on the poles, which are connected externally via the terminal blocks.

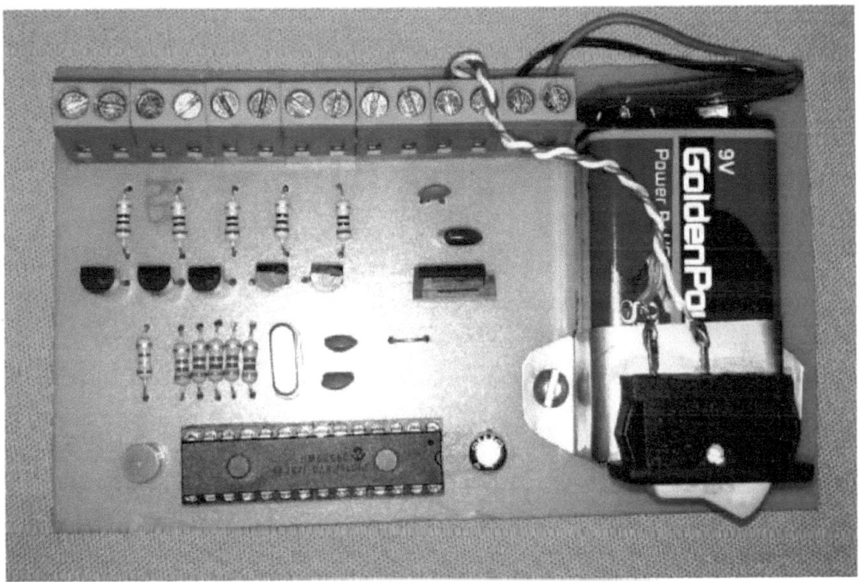

Figure 21 - Physical prototype of the transmitter plant

Source: Author

Figure 22 shows the final circuit of the receiver.

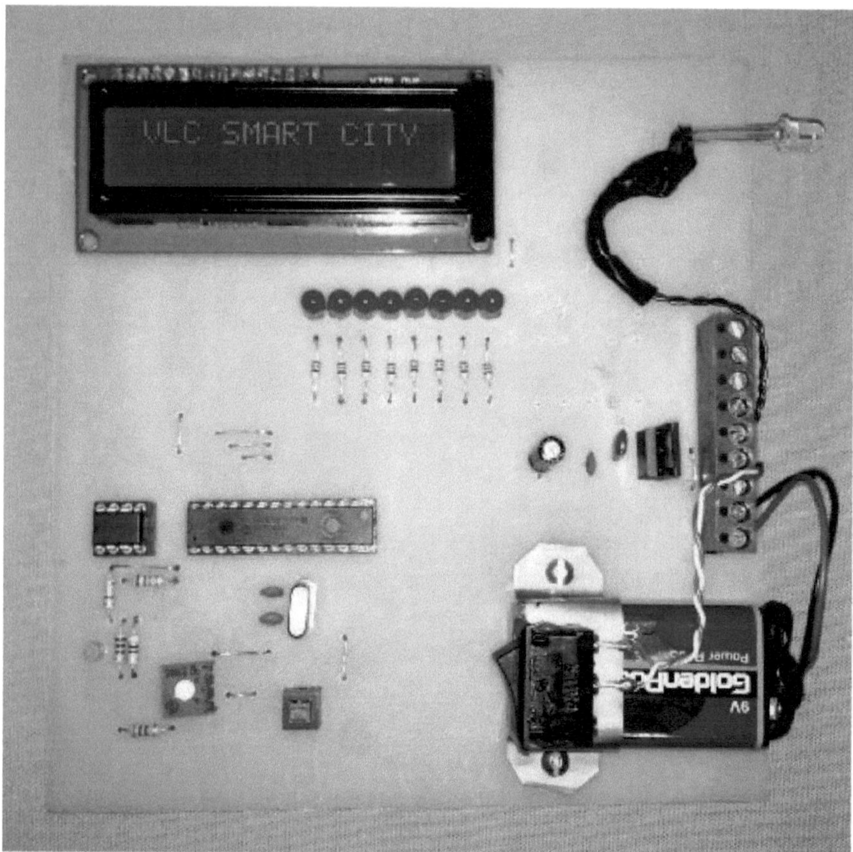

Figure 22 - Physical prototype of the receiver circuit

Source: Author

4.1.2 Communication with Poles and Traffic Lights

Figure 23 to Figure 28 show the communications between the posts/semaphores and the receiver. In each figure, an LED was connected to the post of the respective pole. These pictures were taken for a distance of 30cm between transmitter and receiver, in order to only show the messages being received correctly, so that it would be possible to display both the transmitter and the receiver in the picture. However, the circuit worked up to a distance of 18m and details of operation at maximum range are shown in the quantitative results.

Figure 23 - Photo of the circuit in communication with pole 1

Source: Author

Figure 24 - Photo of the circuit in communication with pole 2

Source: Author

Figure 25 - Photo of the circuit in communication with pole 3
Source: Author

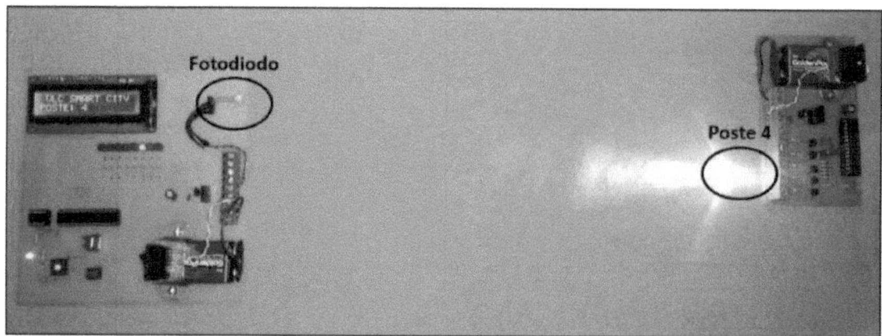

Figure 26 - Photo of the circuit in communication with pole 4
Source: Author

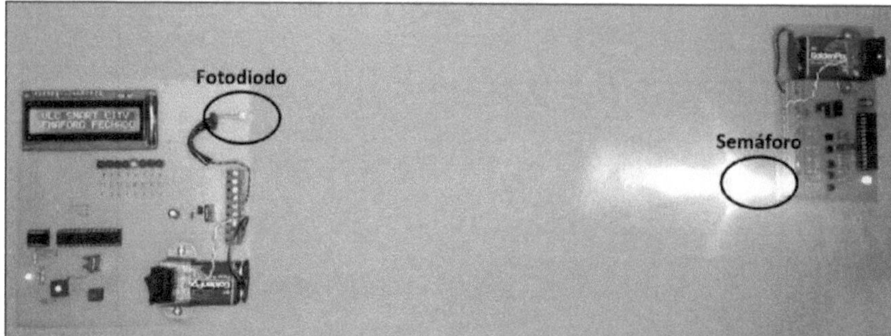

Figure 27 - Photo of the circuit in communication with the closed traffic light
Source: Author

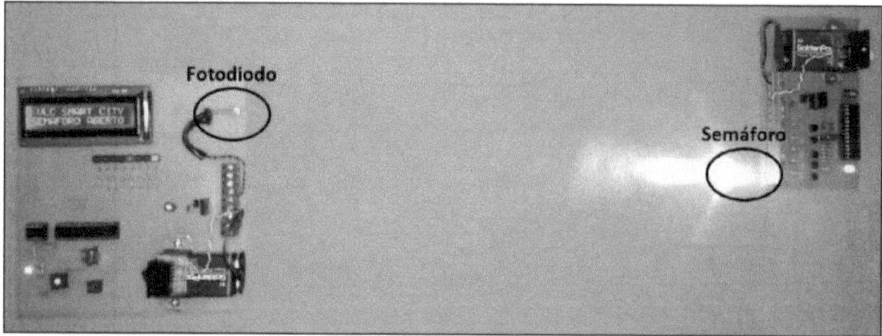

Figure 28 - Photo of the circuit in communication with the open traffic light
Source: Author

From these qualitative results, a good stability in communication can be observed. In the qualitative tests, it was realised that the phototransistor does not need to be fully aligned with the LED of the pole, it is enough to be in front of the emitted light. In the qualitative tests, several ambient lights were purposely placed to cause interference, and the system worked correctly both day and night and not even the camera *flash* when taking the picture was able to impair communication.

4.1.3 Experiments Using Light Amplifying Lenses

Figure 29 shows the transmitter circuit connected to an LED using a focussing lens.

Figure 29 - Transmitter connected to an LED with focusing lens
Source: Author

Figure 30 shows another communication between the receiver and the pole 1 at 2m, without focusing lens. And Figure 31 shows another communication between the receiver and the traffic light open at 8m, with focusing lens. These experiments were carried out at night in an environment lit by a 7W LED lamp.

Figure 30 - Additional photo of the circuit in communication with pole 1
Source: Author

Figure 31 - Additional photo of the circuit in communication with the open traffic light
Source: Author

4.2 Quantitative Results

4.2.1 Maximum Measurements Obtained

The data transmission frequency was calculated, and tests were performed to measure the maximum range of the implemented prototype, the maximum angle of inclination of the phototransistor with respect to the emitting source (angle 0 indicates that the phototransistor is fully aligned with the LED), the maximum displacement of the phototransistor with respect to the emitting source (displacement of 0cm indicates that the phototransistor is fully aligned with the LED), and the maximum speed of the receiver in a moving device, at which data is still received. In this section these maximum values obtained are presented.

Figure 32 illustrates the names of each parameter used in this section.

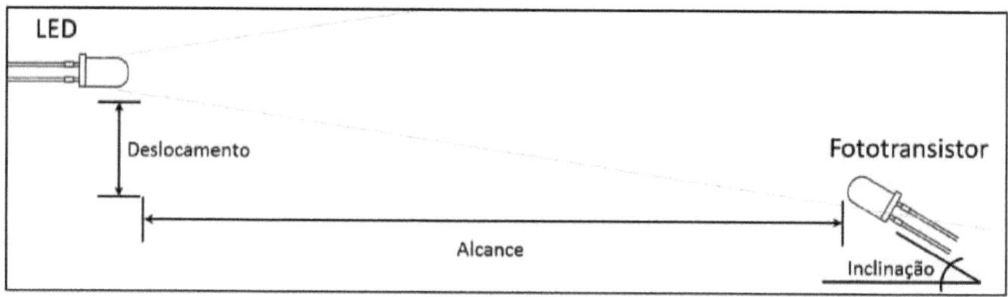

Figure 32 - Illustration of measured parameters

Source: Author

4.2.1.1 Data Transmission Frequency

In the configuration of the UART channels, several bit transfer rates were tested. The rate at which the circuit worked best was 1800bps (CCS allows the rate to be set to values other than those typically standardised in software). When using speeds below this, it was possible to notice the LED flashing with the naked eye. When using speeds above this, the error rates increased significantly.

The rate of 1800 bits per second was therefore chosen, corresponding to 225bytes per second. As all the information is transmitted in a single byte, the data transmission speed is therefore 225 packets per second, equivalent to a transmission frequency of 225Hz.

4.2.1.2 Maximum Range

Table 2 shows the maximum ranges of the circuit, taking into account some significant aspects for the range value. During the day, the tests were carried out in sunlight, making adjustments so that sunlight does not interfere so much in the operation. During the night the tests were done in front of ambient LED lights and no adjustments were necessary. The focussing lens was placed on the emitting LED to concentrate the emitted light, which increased the operating distance but reduced the light detection area by the receiver.

Table 2 - Maximum ranges of the implemented prototype

Condition	Maximum range
By day without focusing lens	2.2m
By day with focusing lens	7.3m
At night without focusing lens	6.3m
At night with focusing lens	18.1m

Source: Author.

In addition to the range of the circuit in communication, the range until the phototransistor is fully open was measured. Without focusing lens the maximum value achieved was 7.7m and with focusing lens the maximum value was 19.4m.

4.2.1.1 Maximum Phototransistor Tilt

The measurement of the phototransistor tilt angle was done using trigonometry and the approximate values are shown in Table 3. The maximum inclinations were measured for some distances at which the system operation was stable and also for distances close to the maximum ranges. At the maximum ranges, the slope is practically zero because it is the limit point of light detection of the LED, and in this, the LED had to be fully aligned with the phototransistor.

Table 3 - Maximum phototransistor slopes for different situations

Condition	Reach	Maximum inclination
By day without focusing lens	0.5m	9.2 degrees
By day without focusing lens	1.0m	12.2 degrees
By day without focusing lens	2.0m	13.1 degrees
By day with focusing lens	2.0m	12.6 degrees
By day with focusing lens	4.0m	14.4 degrees
By day with focusing lens	6.0m	9.8 degrees
At night without focusing lens	2.0m	12.3 degrees
At night without focusing lens	4.0m	13.2 degrees
At night without focusing lens	6.0m	15.4 degrees
At night with focusing lens	6.0m	18.7 degrees
At night with focusing lens	10.0m	16.5 degrees
At night with focusing lens	14.0m	9.4 degrees

Source: Author.

4.2.1.2 Maximum Phototransistor Offset

The measurement of the phototransistor displacement (horizontal distance of the phototransistor from the LED), was done for the maximum range values, and the results are presented in Table 4.

Table 4 - Maximum phototransistor displacements for different situations

Condition	Reach	Maximum displacement
By day without focusing lens	0.5m	31cm
By day without focusing lens	1.0m	23cm
By day without focusing lens	2.0m	21cm
By day with focusing lens	2.0m	19cm
By day with focusing lens	4.0m	33cm
By day with focusing lens	6.0m	21cm
At night without focusing lens	2.0m	22cm

At night without focusing lens	4.0m	37cm
At night without focusing lens	6.0m	25cm
At night with focusing lens	6.0m	36cm
At night with focusing lens	10.0m	48cm
At night with focusing lens	14.0m	39cm

Source: Author.

4.2.1.3 Maximum Receiver Speed in Motion

The speed of the moving receiver was obtained by measuring the time it takes to get from a starting point to an end point, and calculating the speed by the equation speed=distance/time. The calculation was approximate and done for speeds below 30km/h. Table 5 shows the (approximate) maximum speeds at which the receiver can still receive the message, day and night.

Table 5 - Maximum receiver speeds for different situations

Condition	Reach	Maximum speed
By day without focusing lens	2m	23.7km/h
By day with focusing lens	4m	24.2km/h
At night without focusing lens	4m	22.3km/h
At night with focusing lens	8m	25.0Km/h

Source: Author.

4.2.2 Calculation of BER and PER rates

To validate the proposed system, the bit error rate (BER) was calculated for different distances between the LEDs and the phototransistor, leaving it in ambient situations with light interference and movement of the phototransistor. The way the system was designed both in the physical circuit and in the programming of the microcontrollers, led to a system that practically never receives wrong bits, as the system either receives all the bits correctly, or simply does not receive them. The calculation of this BER rate is done by Equation 2.1.

Figure 33 shows the BER values for the different distances between the LEDs and the phototransistor in two different graphs, one for the measurements made during the day and one during the night, in situations where the packet is received.

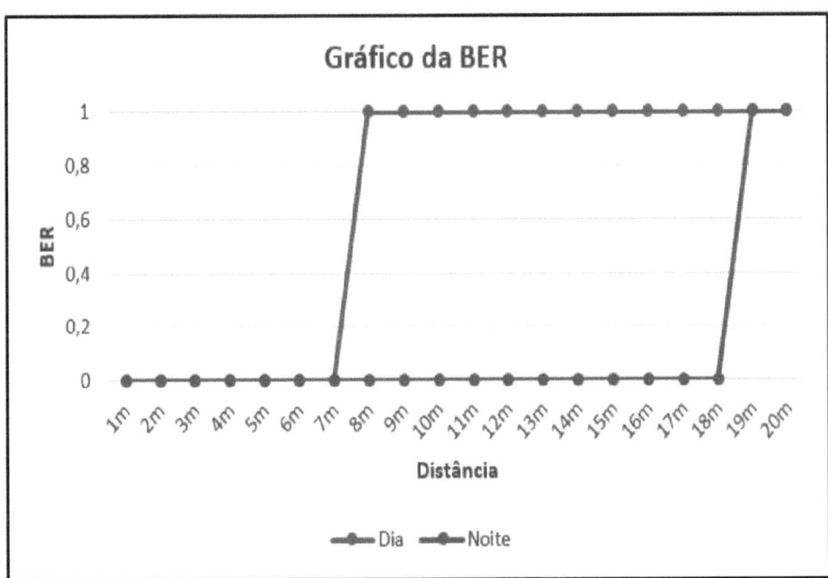

Figura 33   - BER plot for different distances between an LED and the phototransistor

Source: Author

In addition to the BER, to better present the quality of the designed system, the wrong packet rate (PER) was calculated, which consists of what percentage of the packets sent by the transmitter are not correctly received by the receiver in situations with a lot of interference. For this, it was calculated in the transmitter, how many packets are sent per minute and then it was measured in the receiver how many packets it receives per minute, for different distances between the LEDs and the phototransistor in situations with enough interference from ambient light and movement of the phototransistor. The calculation of this PER rate is done by Equation 2.2.

Figure 34 shows the PER values for the different distances between the LEDs and the phototransistor in two different graphs, one for the measurements made during the day and the other during the night. It can be observed that as the distance between the LED and the phototransistor increases, the error rate increases, both during the day and at night, because the greater the distance, the more packets tend to be lost due to interference. However, this is not problematic for this system as in it the information from each pole is transmitted in a single 8-bit packet.

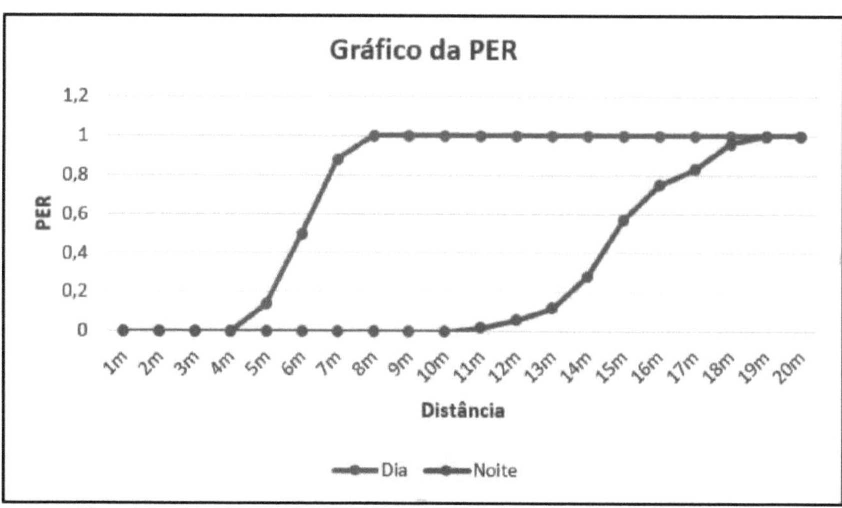

Figura 34   - Graph of PER for different distances between an LED and the phototransistor

Source: Author

The quantitative results validated the effectiveness of the designed system. The main objective of the project was to make it work at a scale of 1:10, but it can be seen from these results that it worked, at excellent distances, even using low power components. For a real situation, which is not included in the objectives of this prototype, distances of up to 50m can be achieved, based on the theoretical framework presented, with greater inclinations and displacements, using higher power LED lamps and phototransistors with greater sensitivity and greater opening angle.

CHAPTER 5

Final considerations

5.1 Conclusions

This work brought a new approach to VLC technology in smart cities that, through a low complexity system, will allow the localisation and decision making of smart vehicles in front of smart poles and traffic lights, reusing the lamps and other resources already existing in the city infrastructure. The results presented showed the quality of the developed system, taking into account that low-power components were used and that it was tested with many ambient interferences, achieving optimal operating distances.

In addition to being expandable to be applied in larger cities with more poles and traffic lights, the prototype presented serves as an excellent basis for other work with VLC that seeks a more efficient strategy that does not require expensive and complex devices to implement.

5.2 Future Work

For future work, one could also detect the power of the LEDs, calculating the intensity of the emitted light, through the receiver to estimate the most accurate position of the vehicles (the closer the vehicle is to the light emitting source, the more intense the light will be received instantaneously) thus increasing the accuracy of the positioning system. In addition, the maximum communication distance between the LEDs and the photodiode can be increased by using higher power components.

References

ANTONIO, M. *Programming PIC Microcontrollers using C language*. [S.l.]: Federal Centre for Technological Education of Espírito Santo, 2006. Cited 4 times on pages 22, 23, 24 and 25.

BOLLIER, D. How smart growth can stop sprawl. *Essential Books*, 1998. Quoted on page 12.

BOUBAKRI, W.; ABDALLAH, W.; BOUDRIGA, N. A light-based communication architecture for smart city applications. *Transparent Optical Networks (ICTON)*, 2015. Cited 2 times on pages 19 and 20.

BOYLESTAD, R. L.; NASHELSKY, L. *Electronic Devices and Circuit Theory*. 8. ed. [S.l.]: Prentice Hall, 2004. Cited 2 times on pages 34 and 37.

CALDWELL, R. *Portland, a city of 'smart growth'*. [S.l.]: The Masthead, 2002. v. 54. cited on page 12.

CANZIAN, E. Serial communication - rs232. *CNZ Engenharia e Informática Ltda*, 2013. Cited 3 times on pages 26, 27 and 28.

CARVALHO, E. A. de; ARAUJO, P. C. de. Basics of global positioning system gps. *Federal University of Rio Grande do Norte*, 2009. Cited 2 times on pages 28 and 29.

HARRISON, C.; DONNELLY, I. A. A theory of smart cities. *55th Annual Meeting of the ISSS*, 2011. Cited 2 times on pages 12 and 13.

KHAN, L. U. Visible light communication: applications, architecture, standardisation and research challenges. *Digital Communications and Networks 3,*, p. 78-88, 2017. Cited 3 times on pages 14, 16 and 21.

KUMAR, N. et al. Visible light communications in intelligent transportation systems. *Intelligent Vehicles Symposium,*, p. 748-753, 2012. Cited 4 times on pages 9, 18, 19 and 30.

KUMAR, N.; NERO, L. A.; AGUIAR, R. L. Visible light communication for advanced driver assistant systems. *Institute of Telecommunications*, 2007. Cited 2 times on pages 9 and 19.

LIU, H. M. X.; MAEDA, Y. Basic study on indoor location estimation using visible light communication platform. 30th Annual International IEEE EMBS Conference, 2008. Cited 2 times on pages 17 and 18.

OWEN, D. Green metropolis. *Riverhead, London,* 2009. Quoted on page 12.

PENIDO, E. C. C.; TRINDADE, R. S. *Microcontrollers*. [S.l.]: Federal Institute of Education, Science and Technology, 2013. Cited 2 times on pages 21 and 22.

PEREIRA, F. *Microcontroladores PIC - Programação em C*. [S.l.]: Editora Érica Ltda, 2003. Cited on page 26.

POHLMANN, C. Visible light communication. *Seminar Kommunikations standards in der Medizintechnik*, 2010. Cited 4 times on pages 9, 10, 14 and 17.

POOLE, I. *BER Bit Error Rate Tutorial and Definition*. 2015. Available at: <http://www. radio-electronics.com/info/rf-technology-design/ber/bit-error-rate-tutorial-definition.php>. Accessed on: 2018-05-20. Cited on page 29.

PREMACHANDRA, H. C. N. et al. High-speed-camera image processing based led traffic light detection for road-to-vehicle visible light communication. *IEEE Intelligent Vehicles Symposium,*, p. 793-798, 2010. Cited on page 19.

ROCHA, J.; GAMEIRO, A. Synchronous and asynchronous data communication systems. *University of Aveiro*, 2015. Cited 2 times on pages 27 and 28.

TONON, R. *Smart Cities*. 2013. Available at: <http://revistagalileu.globo.com/Magazine/Common/0,,ERT338454-17773,00.html>. Accessed on: 2018-01-16. Cited on page 10.

TRUZZI, S. Visible light communication. *Università Degli Studi di Torino*, 2016. Cited 3 times on pages 14, 15 and 17.

YAQOOB, I. et al. Enabling communication technologies for smart cities. *IEEE Communications Magazine,*, p. 112-120, 2017. Cited on page 9.

A Materials Used

A.1 Materials Used in the Transmitter

Table 6 shows the list of materials used to build the prototype transmitter.

Table 6 - List of materials used in the transmitter

Material	Quantity
Voltage regulator IC LM7805	1
100nF ceramic capacitor	2
100uF electrolytic capacitor	1
47pF ceramic capacitor	2
4.43MHz crystal	1
Integrated circuit PIC16F870	1
28-pin turned socket for PICs	2
180ohm resistor	1
56ohm resistor	5
8.2Kohm resistor	5
Transistor BC548	5
Green diffused led 5mm	1
5mm white high-brightness LED	5
On/off toggle switch	1
9V battery	1
9V battery connector	1
9V battery holder	1
Green two-way terminal	7
Single sided phenolite board 11x7cm	1
Various wires for connection	-

Source: Author.

A.2 Materials Used in the Receiver

Table 7 shows the list of materials used to build the prototype receiver.

Table 7 - List of materials used in the receiver

Material	Quantity
Voltage regulator IC LM7805	1
100nF ceramic capacitor	2
100uF electrolytic capacitor	1
47pF ceramic capacitor	2
4.43MHz crystal	1
Integrated circuit PIC16F870	1
28-pin turned socket for PICs	2
LM358 integrated circuit	1
Turned 8-pin socket for LM358	2
150ohm resistor	8
180ohm resistor	1
220ohm resistor	1
1Kohm resistor	2
4.7Kohm resistor	1
8.2Kohm resistor	1
10Kohm resistor	5
22Kohm resistor	1
100Kohm trimpot	1
Zenner diode 1N4728A	5
Green diffused led 5mm	1
Red diffused led 5mm	8
16x2 LCD display	1
Phototransistor receiver 5mm TIL78	1
Push Button Spanner 6 Pins 6mm w/lock	1
On/off toggle switch	1
9V battery	1
9V battery connector	1
9V battery holder	1
Green two-way terminal	6
Single sided phenolite board 15x15cm	1
Various wires for connection	-

Source: Author.

B Implemented Codes

B.1 Microcontroller code of the Transmitter Circuit

```
// TRANSMISSOR
//---------------------------------------------------------------
// Configuracoes

#include <16f870.h>

#FUSES HS, NOWDT, NOPUT, NOBROWNOUT, NOLVP

#USE delay(clock=4433190)

#USE rs232(baud=1800, xmit=PIN_C0, rcv=PIN_C7,
           bits=8, parity=N, errors, stream=chan1)
#USE rs232(baud=1800, xmit=PIN_C1, rcv=PIN_C7,
           bits=8, parity=N, errors, stream=chan2)
#USE rs232(baud=1800, xmit=PIN_C2, rcv=PIN_C7,
           bits=8, parity=N, errors, stream=chan3)
#USE rs232(baud=1800, xmit=PIN_C3, rcv=PIN_C7,
           bits=8, parity=N, errors, stream=chan4)
#USE rs232(baud=1800, xmit=PIN_C4, rcv=PIN_C7,
           bits=8, parity=N, errors, stream=chan5)

/* Obs.: As linhas #USE acima foram quebradas para caberem no arquivo,
mas no codigo nao eh permitida essa quebra de linha. */

//---------------------------------------------------------------
// Enderecos

#BYTE PORTA=0x05
#BYTE PORTB=0x06
#BYTE PORTC=0x07

#BYTE TRISA=0x85
#BYTE TRISB=0x86
#BYTE TRISC=0x87

#BIT LED1=PORTB.0

//---------------------------------------------------------------
// Variaveis globais
```

```c
long int clksem1i=0;
long int clksem1x=500;
int clkled1i=0;
int clkled1x=10;

char poste1=0b11111110;
char poste2=0b11111101;
char poste3=0b11111100;
char poste4=0b11111011;
char semaf1=0b11110111;

//----------------------------------------------------------------
// Programa principal

int main()
{
        TRISB=0x00;

        while(1)
        {
                clkled1i+=1;
                if(clkled1i>clkled1x)
                {
                        clkled1i=0;
                        LED1=~LED1;
                }

                clksem1i+=1;
                if(clksem1i>clksem1x)
                {
                        clksem1i=0;
                        if(semaf1==0b11110111)
                                semaf1=0b11110110;
                        else
                                semaf1=0b11110111;
                }

                putc(poste1,chan1);
                putc(poste2,chan2);
                putc(poste3,chan3);
                putc(poste4,chan4);
                putc(semaf1,chan5);
        }
}
```

B.2 Microcontroller code of the receiver circuit

```c
//RECEPTOR
//----------------------------------------------------------------
//Configuracoes

#include <16f870.h>
#include <string.h>

#FUSES HS, NOWDT, NOPUT, NOBROWNOUT, NOLVP

#USE delay(clock=4433190)

#USE rs232(baud=1800, xmit=PIN_A1, rcv=PIN_A0,
           bits=8, parity=N, errors, stream=chan1)

/*Obs.: A linha #USE acima foi quebrada para caber no arquivo,
mas no codigo nao eh permitida essa quebra de linha.*/

//----------------------------------------------------------------
//Enderecos

#BYTE PORTA=0x05
#BYTE PORTB=0x06
#BYTE PORTC=0x07

#BYTE TRISA=0x85
#BYTE TRISB=0x86
#BYTE TRISC=0x87

#BIT   RS=PORTB.5
#BIT   RW=PORTB.6
#BIT   EN=PORTB.4

#BIT   S1=PORTA.2

//----------------------------------------------------------------
//Variaveis globais

char txt1[16];   //Texto da linha 1 do display
char txt2[16];   //Texto da linha 2 do display

//----------------------------------------------------------------
//Funcoes para LCD

void inst(char v)
{
        PORTB=((v>>4)&0x0F);
```

```c
        RS=0;
        EN=1;
        delay_us(500);
        EN=0;
        PORTB=(v&0x0F)|0x00;
        EN=1;
        delay_us(500);
        EN=0;
}

void dado(char v)
{
        PORTB=((v>>4)&0x0F);
        RS=1;
        EN=1;
        delay_us(500);
        EN=0;
        PORTB=(v&0x0F)|0x20;
        EN=1;
        delay_us(500);
        EN=0;
}

void send_byte(char v, int t)
{
                if(t==0){inst(v);}
        else if(t==1){dado(v);}
}

void lcd_init()
{
        RS=0;RW=0;EN=0;

        send_byte(0x03,0);
        send_byte(0x03,0);
        send_byte(0x03,0);
        send_byte(0x02,0);
        send_byte(0x28,0);
        send_byte(0x0C,0);
        send_byte(0x06,0);
        send_byte(0x80,0);
}

void lcd_gotoxy(char x, char y)
{
    char address;
```

```c
   if(y!=1)
     address=0x40;
   else
     address=0;
   address+=x-1;
   send_byte(0x80|address,0);
}

void lcd_putc(char c)
{
 switch(c)
   {
     case '\f':
       send_byte(1,0);
       delay_ms(2);
       break;

     case '\n':
        lcd_gotoxy(1,2);
        break;

     case '\b':
        send_byte(0x10,0);
        break;

     default:
        send_byte(c,1);
        break;
   }
}

void lcd_txt(char t1[])
{
        char i;
        for(i=0;t1[i]!='\0';i++)
        {
                lcd_putc(t1[i]);
        }
}

void lcd_int(long val)
{
        char symb[11] = "0123456789";
        long v1=0,div=1;

        if(val<10){lcd_putc(symb[val]);}
        else
```

```c
        {
                while ( div < val )  div *= 10;
                while ( div > 1)
                {
                        v1 = ( val%div )/( div /10 );
                        if ( div==val ){ v1 =1; }
                        lcd_putc (symb[ v1 ] );
                        if ( div==val ){ lcd_putc (symb[0] ); }
                        div /= 10;

                }

        }
}


void lcd_txtint ( char t1 [] , long val )
{
        lcd_txt ( t1 );
        if ( val <0){ val *=-1;lcd_putc ( '-' ); }
        lcd_int ( val );
}


//————————————————————————————————————————————————————————————————
// Funcao de Inicializacao

void init ()
{
        TRISA=0xFF;
        TRISB=0x00;
        TRISC=0x00;

        PORTB=0x00;
        PORTC=0x00;

        lcd_init ();
        strcpy ( txt1 , "_VLC_SMART_CITY__" );
        lcd_gotoxy (1 ,1);
        lcd_txt ( txt1 );
}


//————————————————————————————————————————————————————————————————
// Programa principal

void main ()
{
        char dattmp1 =0;
        char dattmp2 =0;
        char datget =0;
        int tmg1i =0;
```

```c
int  tmg1x=1;

int  tmd1i=0;
int  tmd1x=20;

int  tml1i=5;

int  vtmp1=0;

init ();

while(true)
{
        if(tml1i==0)
                dattmp1=~getc(chan1);
        else
                tml1i -=1;

        if(dattmp2==dattmp1)
        {
                tmg1i+=1;
                if(tmg1i>tmg1x)
                {
                        datget=dattmp1;
                }
        }
        else
        {
                tmg1i=0;
                datget=0b00000000;
                dattmp2=dattmp1;
        }

        vtmp1=0;

        if(datget!=0b00000000)
        {
                if(datget==0b00001000)
                {
                        PORTC=datget;
                        strcpy(txt2 , "SEMAFORO_FECHADO");
                        lcd_gotoxy(1,2);
                        lcd_txt(txt2);
                }
                else if(datget==0b00001001)
                {
                        PORTC=datget;
```

```c
                strcpy(txt2,"SEMAFORO_ABERTO_");
                lcd_gotoxy(1,2);
                lcd_txt(txt2);
            }
        else if(datget>=1 && datget<=4)
        {
                PORTC=datget;
                strcpy(txt2,"POSTE:_");
                lcd_gotoxy(1,2);
                lcd_txtint(txt2,datget);
                strcpy(txt2,"__________");
                lcd_txt(txt2);
        }
        else
        {
                vtmp1=1;
        }
    }
    else
    {
        vtmp1=1;
    }

    if(vtmp1==1)
    {
        tmd1i+=1;
        if(tmd1i>tmd1x)
        {
                tmd1i=0;
                strcpy(txt2,"________________");
                lcd_gotoxy(1,2);
                lcd_txt(txt2);
                PORTC=0x00;
        }

        if(datget!=0b11111111)
        {
                tmd1i=0;
        }
    }
  }
}
```

I want morebooks!

Buy your books fast and straightforward online - at one of world's fastest growing online book stores! Environmentally sound due to Print-on-Demand technologies.

Buy your books online at
www.morebooks.shop

Kaufen Sie Ihre Bücher schnell und unkompliziert online – auf einer der am schnellsten wachsenden Buchhandelsplattformen weltweit! Dank Print-On-Demand umwelt- und ressourcenschonend produziert.

Bücher schneller online kaufen
www.morebooks.shop